How the States Got Their Shapes

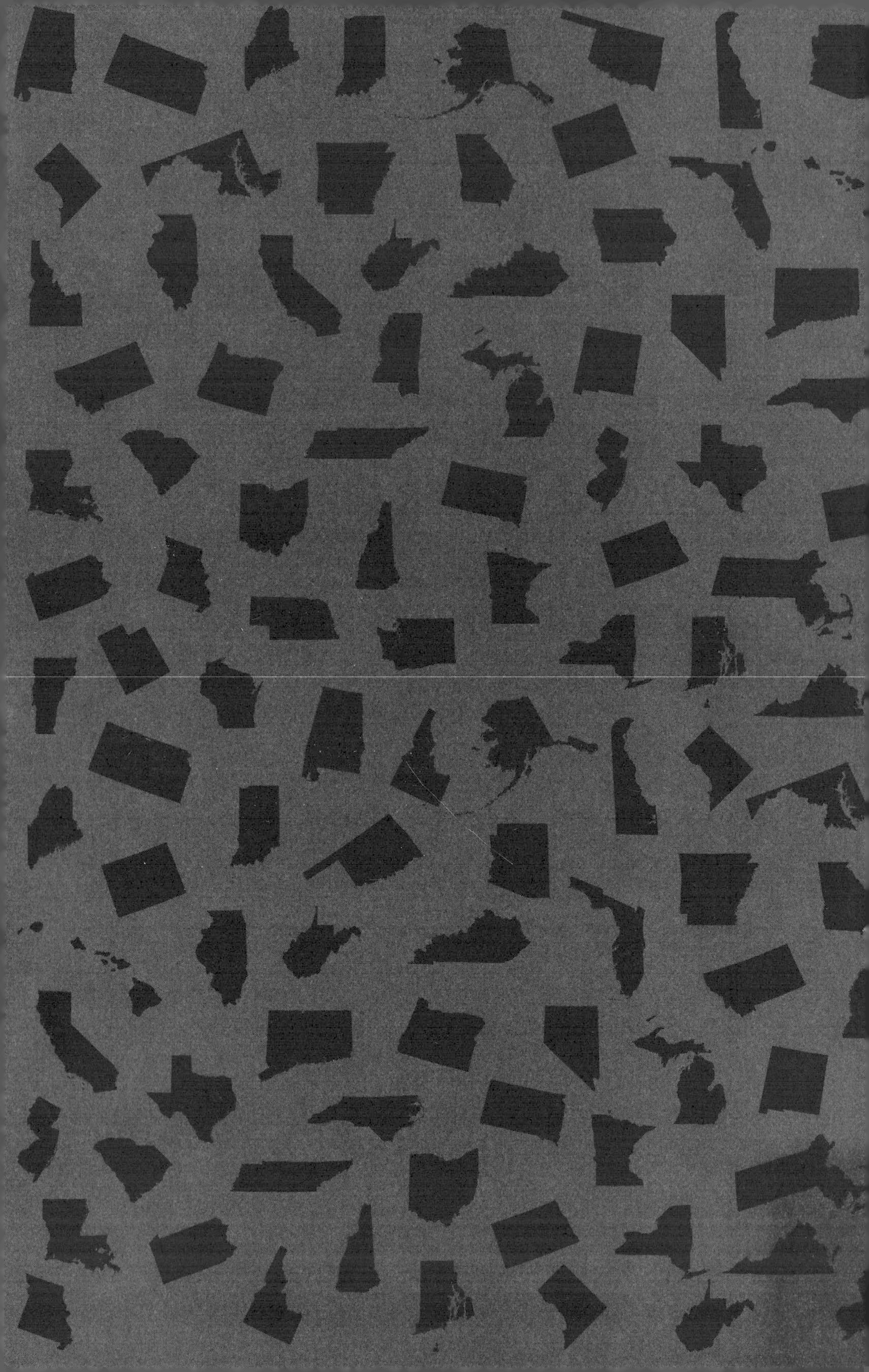

How the States Got Their Shapes

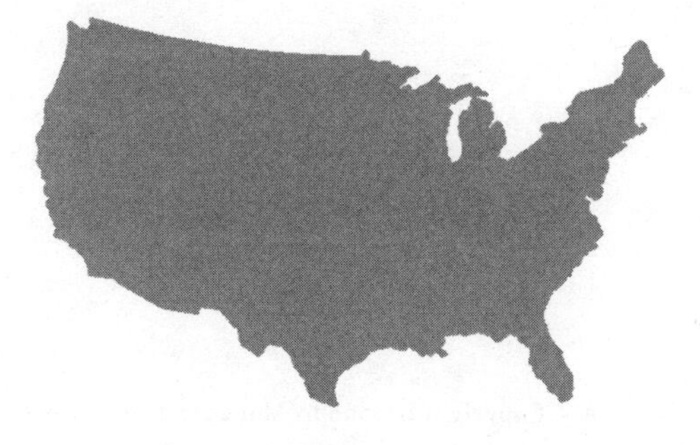

MARK STEIN

Collins

An Imprint of HarperCollins*Publishers*

Published 2008 in the United States of America by Smithsonian Books in association with HarperCollins Publishers.

Designed by Chris Welch.
Maps by XNR Productions, Inc.

ISBN 978-0-06-143138-8

Printed in the U.S.A.

Book Club Edition

Contents

CANADA
Washington
Montana
North Dakota
Missouri
Oregon
Idaho
South Dakota
Wyoming
Nebraska
Nevada
Utah
Colorado
California
Colorado
Kansas
Oklahom
Arizona
New Mexico
PACIFIC OCEAN
Rio Grande
Texa
RUSSIA
ARCTIC OCEAN
Alaska
CANADA
Bering Sea
MEXICO
PACIFIC OCEAN

CANADA
Lake Superior
St. Lawrence
Maine
Vermont
New Hampshire
innesota
Lake Huron
Lake Ontario
Massachusetts
Wisconsin
Lake Michigan
Michigan
New York
Rhode Island
Connecticut
Lake Erie
Pennsylvania
Iowa
New Jersey
Ohio
Maryland
Delaware
Indiana
D.C.
Illinois
West Virginia
Virginia
Ohio
Missouri
Kentucky
North Carolina
Tennessee
South Carolina
Arkansas
Mississippi
Alabama
Georgia
Mississippi
ATLANTIC OCEAN
Louisiana
Florida
Gulf of Mexico
BAHAMAS
Hawaii
CUBA
ACIFIC OCEAN

Acknowledgments

I would like to acknowledge the men and women of the U.S. Geological Survey who created the *National Atlas of the United States.* This interactive Web site has been of immense value to me in preparing this work. Of even greater value has been Caroline Newman, Executive Editor at Smithsonian Books. I now know what authors mean when they praise their editors for having vitally important insight and perseverance. I am also deeply indebted to our copyeditor, Cecilia Hunt, and to our cartographer, Rob McCaleb of XNR Productions, whose keen eyes not only detected errors but also led me to new discoveries. And lastly, my heartfelt thanks to Arlene Balkansky of the Library of Congress, who is my wife. I would urge any young person entering a career that involves research to marry an employee of the Library of Congress.

Introduction

To teach us the boundaries of the states, my seventh grade geography teacher would hold up cutouts and we would raise our hands, vying for the chance to identify which state had the corresponding shape. How we distinguished Wyoming from Colorado, both rectangles, eludes me these many years later. Maybe she just didn't include them. After all, how much value is there in knowing which rectangle is Wyoming and which is Colorado?

Later in life, I came to realize that there is value in learning about the borders of Colorado and Wyoming, but that value resides, not in knowing *what* their shape is, but in knowing *why* it is. Why, for instance, are the straight lines that define Wyoming located where they are and not, say, ten miles farther north or west? Far more knowledge results from exploring why a set of conditions exists than from simply accepting those conditions and committing them to memory. Asking why a state has the borders it does unlocks a history of human struggles—far more history than this book can contain, though this book does aspire to unearth the keys.

Consider for a moment the cluster of states composed of Maine, New Hampshire, and Vermont. Why don't those states extend to their natural boundaries, the St. Lawrence River and the Atlantic Ocean? How come Canada got that land? (Figure 1)

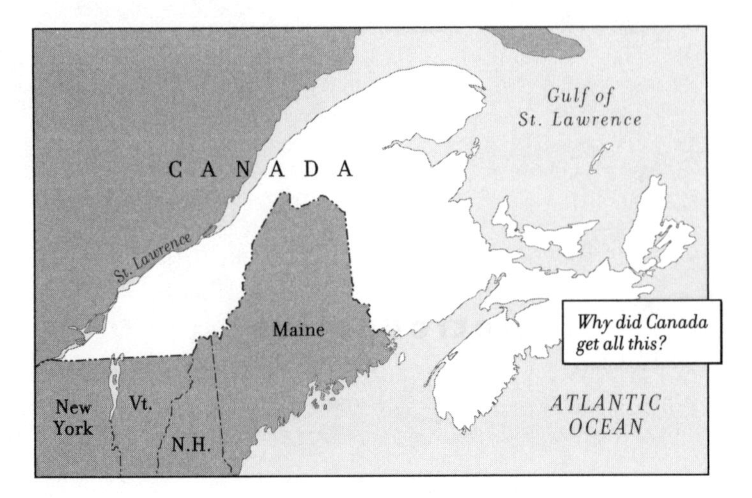

FIG. 1 U.S./Canadian Border

Why does Delaware have a semicircle for its northern border? What's at its center and why was it encircled? Why does Texas have that square part poking up? And why does the square part just miss connecting with Kansas, leaving that little Oklahoma panhandle in between? (Figure 2)

The more one looks at state borders, the more questions those borders generate. Why do the Carolinas and Dakotas have a North state and a

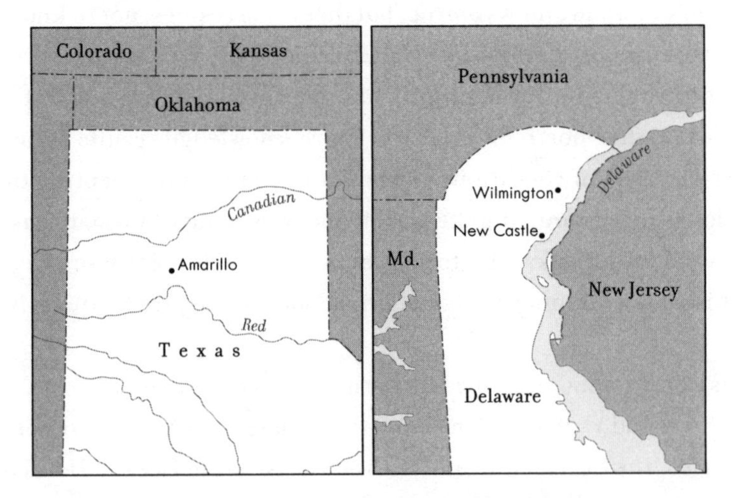

FIG. 2 Border Curiosities: Panhandles and Semicircles

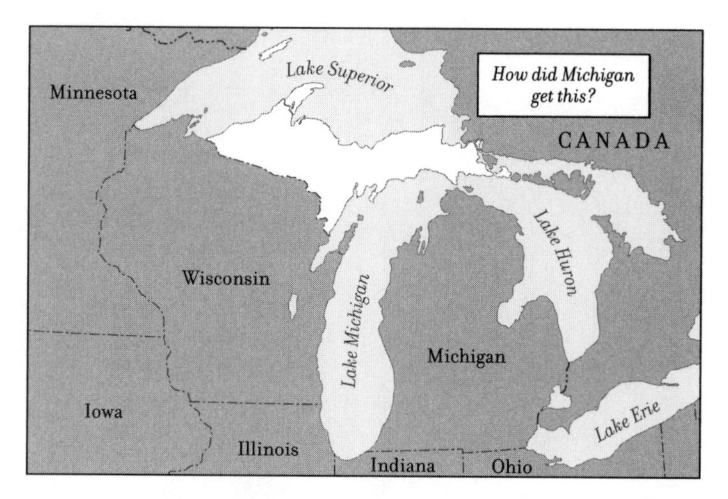

FIG. 3 Wisconsin/Michigan Border

South state? Couldn't they get along? Why is there a West Virginia but not an East Virginia? And why does Michigan have a chunk of land that's so obviously part of Wisconsin? It's not even connected to the rest of Michigan! (Figure 3)

This book will provide those answers. State by state (along with the District of Columbia), the events that resulted in the location of each state's present borders will be identified.

A state border is both an official entrance and a hidden entrance. The official entrance is the legal threshold to a state. But its hidden entrance beckons us to the past. Here at the state line we can come in contact with struggles long forgotten and now overgrown by signs saying things like "Welcome to Kansas—Please Drive Carefully."

Don't Skip This

You'll Just Have to Come Back Later

Many of our state borders are segments of borders that date from England's and, later, the United States' territorial acquisitions, and they can be identified by looking for lines that provide multistate borders.

The French and Indian War Border

The French and Indian War (1754–63) resulted in the oldest of these multistate boundaries. In this war, England and her American colonists began what became the dismantling of France's possessions in North America. With this victory, England added to her North American possessions all the land between the Ohio River and the Mississippi River. The boundaries of that war are still on the map today, for they provide borders for the states of Ohio, Indiana, Illinois, and Wisconsin. (Figure 4)

The division of this land acquired in the French and Indian War influenced virtually every state border that followed. After the Revolution, Congress had to decide how best to divide this region, known as the Northwest Territory, into states. Congress assigned Thomas Jefferson the task of studying this matter and in 1784 Jefferson issued a report to Congress in which he proposed that the region be divided into

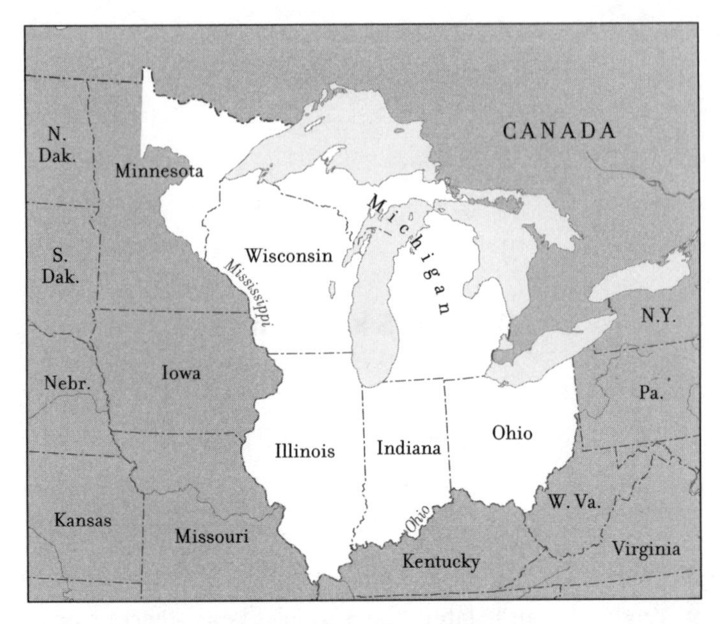

FIG. 4 Land Acquired in the French and Indian War

states having two degrees of height and four degrees of width, wherever possible.

As it turned out, Congress didn't employ these borders when it enacted the Northwest Ordinance in 1787, the law that included the boundary lines for the future states to be created from the Northwest Territory. Congress did, however, adopt its underlying principle: All states should be created equal.

The Louisiana Purchase Borders

Probably the most notable American boundary is the long straight line that defines so much of the nation's northern border with Canada. This line is the 49th parallel. It first surfaced on the American map following the 1803 Louisiana Purchase. The document conveying France's remaining North American land—a tract that included all or some of Louisiana, Arkansas, Oklahoma, Missouri, Kansas, Iowa, Nebraska, South Dakota, Wyoming, Minnesota, Montana, and Colorado—states that the French

Republic cedes to the United States "the Colony or Province of Louisiana with the same extent that it now has." This wording seems refreshingly brief and to the point for a legal document, if a bit vague. The vagueness is also the reason very little evidence of the Louisiana Purchase can be found in our state lines. Other than the boundaries provided by the Gulf of Mexico and the Mississippi River, no one knew what its boundaries were! Jefferson believed that all the land comprising the watershed leading to the Missouri and Mississippi rivers constituted the Louisiana Purchase. But, as he soon discovered, the United States' neighbors did not. In reality, France's American territory extended to the west as far as a Frenchman could go without getting shot by a Spaniard, and likewise to the north without getting shot by an Englishman. (Figure 5)

The ambiguous borders of the Louisiana Purchase led England and the United States to negotiate where France's former lands ended and where British North America (Canada) began. Under the Convention of 1818, the two nations agreed upon the 49th parallel from the westernmost longitude of Lake of the Woods to the crest of the Rocky Mountains. (Figure 6)

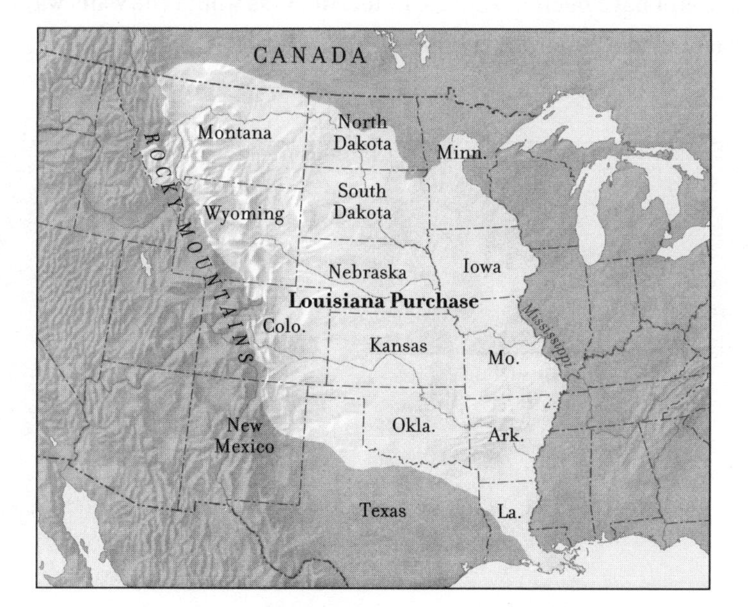

FIG. 5 The Louisiana Purchase—1803

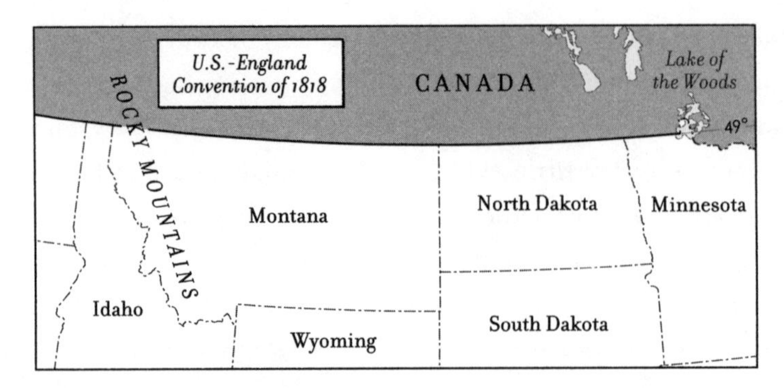

FIG. 6 The First Surfacing of 49°

But the choice of the 49th parallel begs the question, why not 50? It's such a nice round number. The reason for the one-degree difference is that England needed to maintain her access to the Great Lakes via the westernmost of those lakes, Lake Superior. Such access was vital to England's fur trade in general and, in specific, to a major fur trading post located at the confluence of the Assiniboine and Red rivers—a place now known as Winnipeg. Had the border been located at the 50th parallel, Winnipeg would have been in American territory, as would the waterways that flow east to Lake Superior. (Figure 7)

FIG. 7 The Reason for 49° Instead of 50°

The Louisiana Purchase also sparked concern in Spain, which claimed much of the land west of the Rockies. This concern led to the Adams-Onis Treaty (1819). The entire eastern border of Texas—the straight line of what later became its panhandle, the eastward flowing Red River, the straight line southward at the lower corner of Texas, and the Sabine River arcing southward to the Gulf of Mexico—all dates back to this treaty. Also emanating from the Adams-Onis Treaty is the long, multistate line that runs along the 42nd parallel, which later became the northern border of California, Nevada, and northwest Utah. But, as with every man-made line, there is the question, why put the line *there*? (Figure 8)

The Border Inherited from England and Spain

The 42nd parallel already existed as a boundary before the 1803 Louisiana Purchase. In 1790, England and Spain had concluded a treaty known as the Nootka Convention.

The Nootka Convention? Nootka Sound is a small inlet in Vancouver Island (off the west coast of what is now Canada). In the late 1700s, England and Spain nearly went to war over their conflicting ambitions along the west coast of North America. The British needed the rivers and inlets of Vancouver Island and northwestern Canada to carry on their fur trade. But

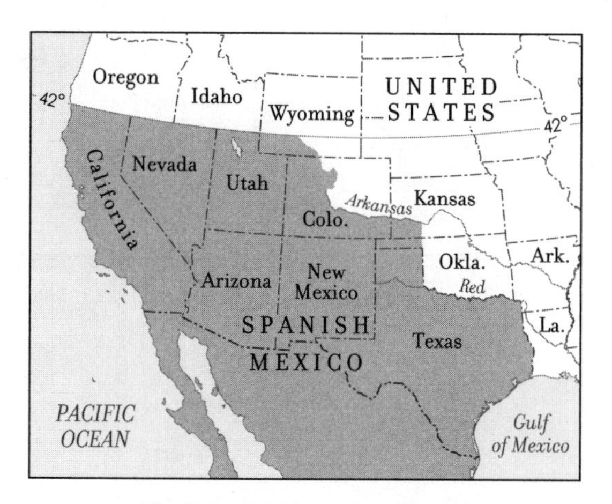

FIG. 8 The Boundary from the Adams-Onis Treaty

the Spanish claimed the land was already theirs—as is reflected to this day by names along the west coast of Canada such as the Juan de Fuca Strait, Port San Juan, Estevan Point, Vargas Island, Valdes Island, and Gabriola.

All of England's interests west of the Rockies would have been at risk had Spain succeeded in ousting England from Vancouver Island. In the Nootka Convention, England sought to protect its access to all those rivers and their tributaries vital to carrying out the fur trade. Key among those rivers was the Columbia River. And virtually all of the tributaries to the Columbia River are north of the 42nd parallel.

Below the 42nd parallel, virtually all of the rivers wend their way to San Francisco Bay. Commerce along these waterways was reserved exclusively for Spain with the 42nd parallel as the border. For its era, it was a great parallel. And evidently it has remained an effective border, since it's still there, dividing five states: Oregon, Idaho, California, Nevada, and Utah. (Figure 9)

Multistate Borders Resulting from Slavery

Not all of our multistate borders have resulted from large land acquisitions. Some multistate lines are vestiges of America's internal affairs.

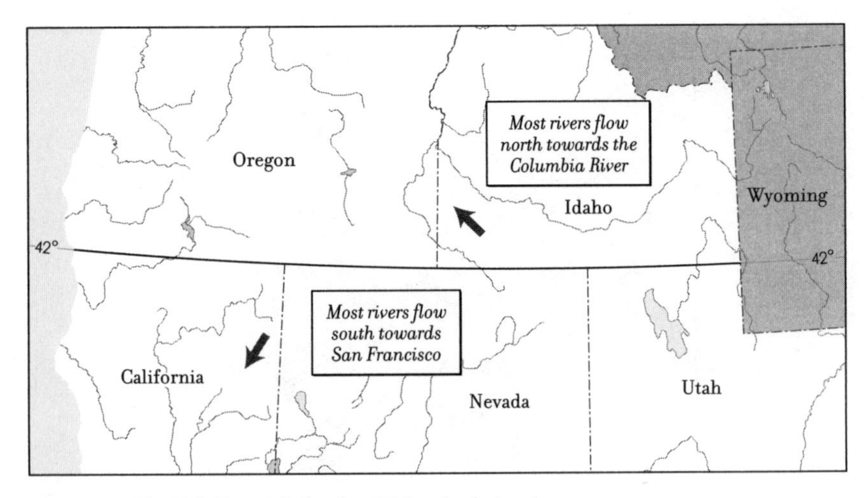

FIG. 9 The U.S./Spanish Border: Watershed at 42°

The long line that defines (with a few deviations) the southern borders of Virginia, Kentucky, Missouri, and the Oklahoma panhandle tells a tale of the struggle to contain the conflict over slavery. Just above that line is another line, this one defining the northern border of the Oklahoma panhandle and the southern borders of Kansas, Colorado, and Utah. The story behind this second line reveals the failure of the struggle to compromise when the issue was slavery. (Figure 10)

Multistate Borders That Do Not Connect

Some of the most revealing multistate borders in the United States are difficult to detect because *they do not connect!* Despite this seeming contradiction, they are indeed multistate borders. Moreover, once detected, they reveal most strikingly the commitment by Congress to the principle that all states should be created equal.

One of these sets of multistate borders consists of the northern and southern borders of Kansas, Nebraska, South Dakota, and North Dakota. Though the borders of each of these states was finalized at a separate time, Congress located each with the result that all four of these prairie states have three degrees of height. (Figure 11)

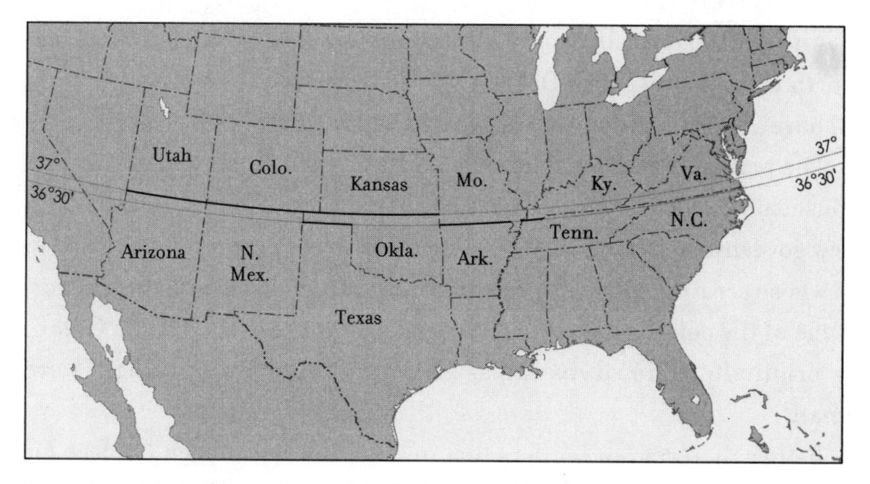

FIG. 10 The Failure to Avoid Civil War Preserved in State Borders

FIG. 11 Prairie States—Equality of Height

Just to the west of this column of states, the Rocky Mountain states of Colorado, Wyoming, and Montana, the borders of which were also created at separate times, all share the fact that they have four degrees of height. (Figure 12)

And the western states of Washington, Oregon, Colorado, Wyoming, North Dakota, and South Dakota, again created over a number of years, all have almost exactly seven degrees of width. (Figure 13)

The principle that all states should be created equal was deeply and consciously rooted in the foundations of the United States. In forming a new government, our founders inherited thirteen colonies, now states, in whose creation by the British crown equality had never been a factor. Some of the colonies were huge, extending—if not to the Pacific Ocean, as originally claimed—as far as the Mississippi River. Others were small.

Rather than redraw borders that in some cases had been in place for as long as 150 years, the founders sought to equalize the inequality of

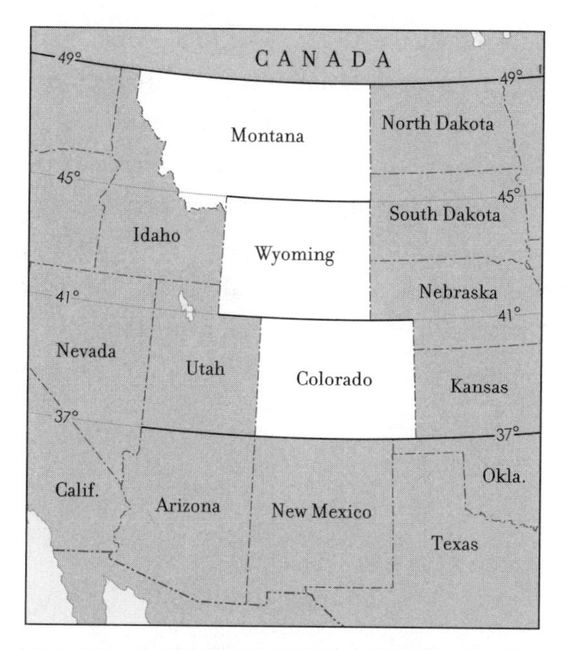

FIG. 12 Rocky Mountain States—Equality of Height

the first thirteen states, in part, by creating a bicameral legislature. The House of Representatives, in which representation is apportioned by population, favored the larger states. But the Senate, in which all states have two votes, favored the smaller states.

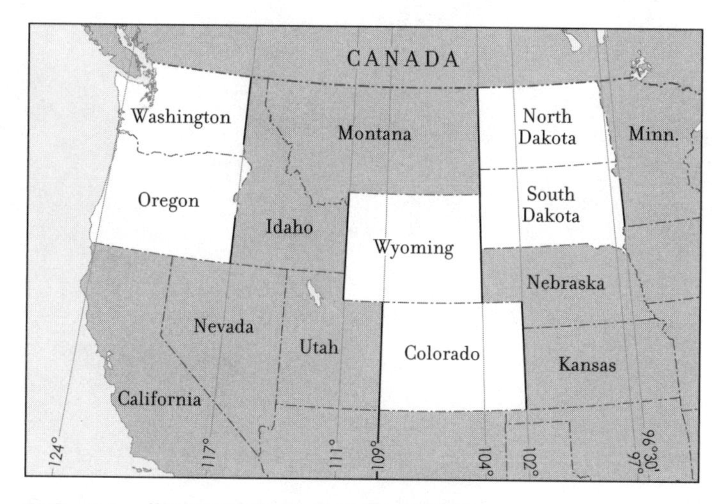

FIG. 13 Western States—Equality of Width

This solution contained a great irony. In a sense, the founders were imitating the government they had just overthrown! England's Parliament is also bicameral, having the House of Commons and the House of Lords. But the House of Lords existed not to create equality but to preserve *inequality*—in the form of the traditional privileges of the nobility.

But before becoming too smug, we need to ask if, in fact, all states are created equal. What about Texas and California? Don't they contradict this principle? Or are they exceptions that prove the rule? To find out, we now need to visit the individual states.

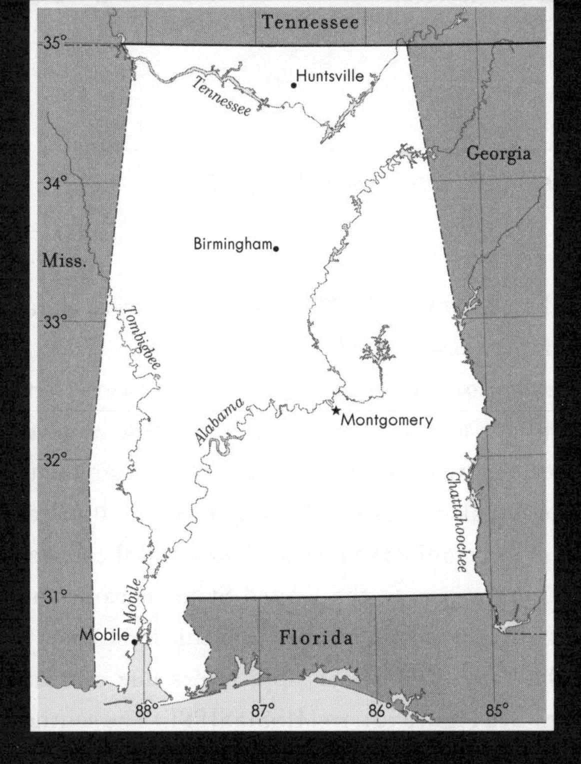

Why does Alabama share a straight-line northern border with Mississippi and Georgia? Why is it almost a mirror image of Mississippi? How come its western side has a tab at the bottom and a little nib at the top? And since Alabama's eastern and western borders have straight-line segments, why don't they go north and south?

The land that is now Alabama was originally part of the Georgia Colony, whose borders extended, according to its royal charter, to the Pacific

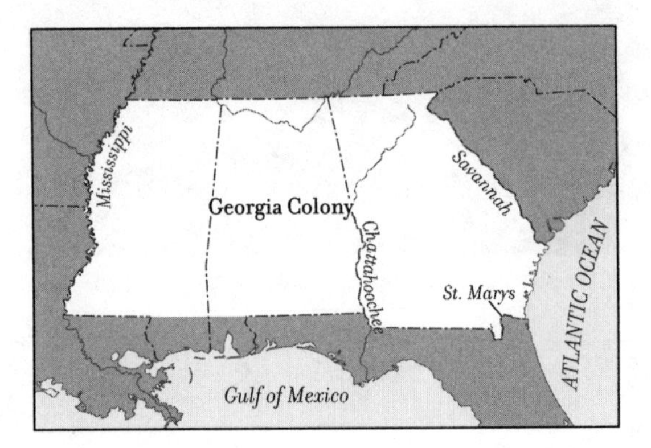

FIG. 14 **The Georgia Colony—1732**

Ocean. In reality, however, the Mississippi River was as far west as the Georgia Colony ever asserted its claim. (Figure 14)

After the Revolution, Congress urged those states with land claims west of the Appalachians to donate that land to the federal government. The plan was to create more states, more equal in size than the thirteen colonies. Since the Appalachians barely enter Georgia, it released its claim instead to any of its colonial lands west of the Chattahoochee River.

The land Georgia gave to the United States became the Mississippi Territory. Initially, the northern border of this land was a line from the juncture of the Yazoo and Mississippi rivers due east to the Chattahoochee River. (See Figure 101, in MISSISSIPPI.) The southern boundary was the 31st parallel. Both these borders had their origins in England's often shifting colonial boundary with Spain. (For details, go to FLORIDA.) But the northern border, short-lived though it was, also reflected a difficult situation between the United States and France. Many Frenchmen lived in the region above the Yazoo and east of the Mississippi. And France had recently been our vital ally in the American Revolution.

Alabama's Northern Border

The Louisiana Purchase, in 1803, changed the equation in terms of any French claims to North American land. Consequently, a new northern border was established for the Mississippi Territory. By extending Georgia's border with North Carolina, Congress fixed a line at 35° as the division between the Mississippi Territory and its northern neighbors, North Carolina and Tennessee. But why 35° in the first place?

The use of this latitude dates back to 1730, when a commission appointed by King George II formulated a border to divide the Carolina Colony into North and South Carolina. (To learn the reasons for 35° in this instance, go to NORTH CAROLINA.) Georgia did not exist until 1732, so for two years the boundary of North and South Carolina, at their western ends, was 35°. When Georgia was created, it inherited that border. The 35th parallel has served as the northern border of Georgia and, by extension, Alabama and Mississippi, ever since.

Though not exactly. Where Alabama's northern border meets Mississippi's northern border, the line jogs slightly to the south. (Figure 19) This adjustment corrected a deviation in the surveying of the western end of Alabama's northern border.

Alabama's Eastern Border

Redefining the northern border of the Mississippi Territory required updating of the territory's eastern border with Georgia (what would become the eastern border of Alabama), since the original border, the Chattahoochee River, turns eastward in this newly added region. At the point where the Chattahoochee turns east, a straight line takes over and heads to the northern border. But while it is a straight line, it is not a north-south line.

The reason for the angle has to do with coal. The straight-line segment of the Alabama/Georgia border heads toward the western range of the Appalachians as they come to their end just below the 35th parallel. Those mountains contained coal, which the Georgians mined up into the 20th century. Had the line gone due north, those mines would have ended up in Alabama. Georgia may have been generous in releasing its western land, but it wasn't that generous.

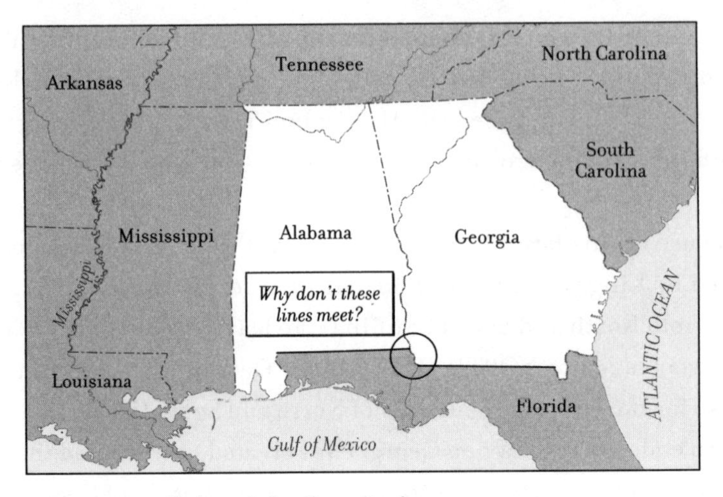

FIG. 15 **Alabama's Southern Border**

Alabama's Southern Border

Since Alabama and Mississippi are an extension of Georgia, why are their southern borders not aligned with that of Georgia? (Figure 15)

In 1739, Spain (which controlled Florida) and the colony of Georgia battled over their border. Georgia emerged the victor and the border that resulted is the border that we see today. It follows the St. Marys River from the coast to its headwaters in the Okefenokee Swamp, and then continues as a straight line westward to the mouth of the Chattahoochee River. (Figure 16)

But west of the Chattahoochee the boundary jumps to 31°—the line stipulated in Georgia's colonial charter. Since Alabama and Mississippi were created from land west of the Chattahoochee River, their southern boundaries remain to this day at 31° while Georgia's southern border remains the line resulting from its 1739 victory over Spain.

The tab extending from the southern border of Alabama reflects the fact that Spanish Florida originally extended all the way around the Gulf of Mexico to the Mississippi River. With the approach and onset of the War of 1812, Spain, no longer the military power it had been, allied itself to British interests. In response, the United States took this opportunity

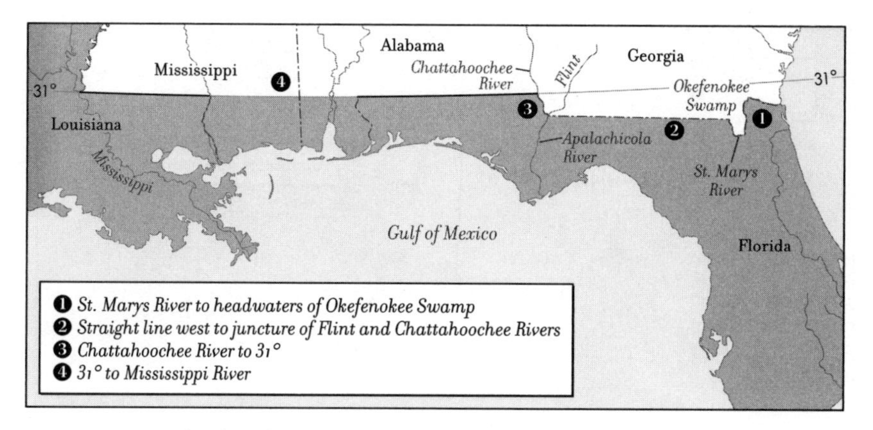

FIG. 16 England/Spain Boundary Agreement—1739

to acquire greater access to the Gulf and control of both banks of the Mississippi by seizing two tracts of Spanish Florida. The first of these it annexed to Louisiana. But it annexed the second tract that it seized to the Mississippi Territory. (See Figure 102, in MISSISSIPPI.) When the Mississippi Territory was divided to create the states of Alabama and Mississippi, this land that had been Florida was split and became the tabs at the bottom of Alabama and Mississippi.

Alabama's Western Border

Alabama came into being in 1816, emerging as a virtual mirror image of Mississippi—the mirror reflecting the belief that all states should be created equal. The Mississippi Territory had been nearly twice as large as its parent, Georgia—a ratio that was not accidental. By dividing the Mississippi Territory in half, three states virtually equal in size were the result.

In addition, the division of the territory was an event its residents welcomed, since it behooved those who favored slavery to create as many slave states as possible, to increase their voting strength in the U.S. Senate.

In dividing the Mississippi Territory, the border that was created to separate Alabama and Mississippi consists of two straight-line

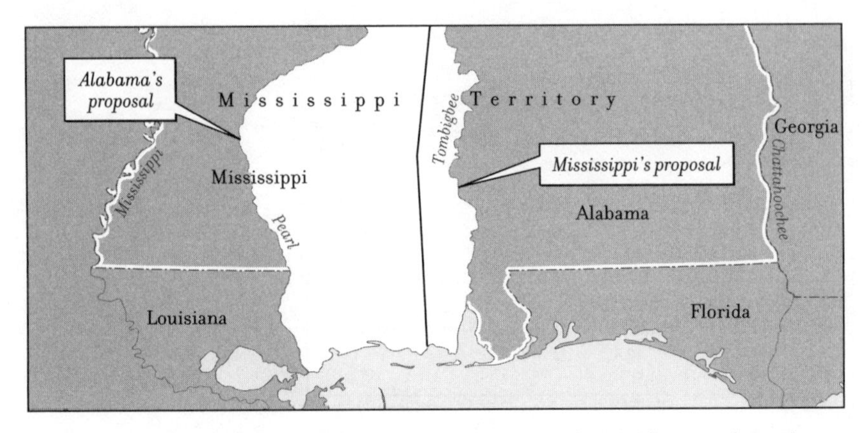

FIG. 17 Mississippi Territory—Conflicting Proposals for Dividing Southern Tier

segments—the lower portion heading north and south, and the upper portion tilting somewhat eastward. Why not just continue the north-and-south portion?

The southern tier of the Mississippi Territory consisted of highly productive bottomland fed by rivers that were themselves highly prized since they flowed directly to the Gulf of Mexico and the sea. When residents in the western half of the territory first contemplated division of the territory for statehood, they proposed the Tombigbee River as the boundary

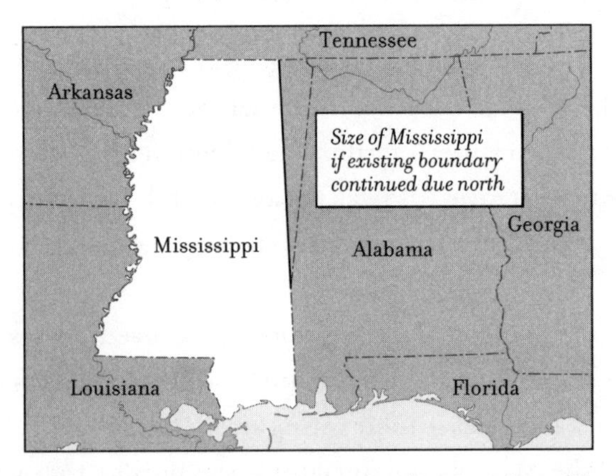

FIG. 18 Alabama's Western Angle—Equality with Mississippi

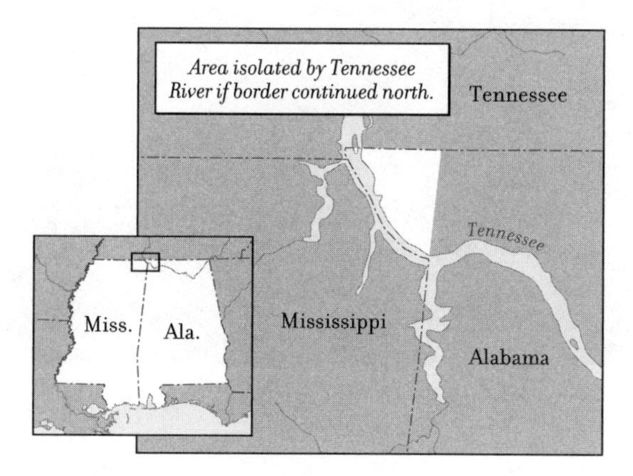

FIG. 19 The Nib in Alabama's Northwest Corner

between Mississippi and Alabama. Residents on the eastern side, however, countered this by proposing the Pearl River as the boundary. Congress opted to divide the territory's rich southern tier equally, establishing a line that extended due south from what was then the northwest corner of Washington County to the Gulf of Mexico. (Figure 17)

If Congress had continued this line to the territory's northern border, Alabama would be considerably larger than Mississippi. (Figure 18) Therefore Congress employed an angled line through the remainder of the territory.

But why does this eastwardly tilted line jog westward at the last minute? (Figure 19) That jog to the northwest is actually the Tennessee River, which takes over as the border at the point where it meets the straight line. Had the straight line continued, Mississippi would have had jurisdiction over a small piece of land cut off from the rest of Mississippi by a wide river in hilly country. In the early 19th century, that was not a good formula for law and order.

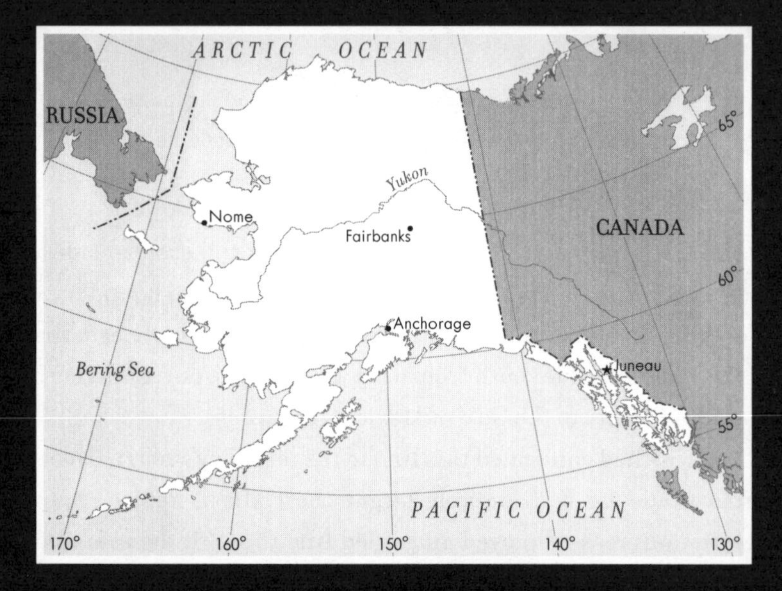

How come Alaska slips out beneath its straight-line eastern border with Canada? In fact, why isn't Alaska just a continuation of Canada? Were the Canadians suckered? Or did we threaten them? And why is Alaska's straight-line border where it is?

In acquiring Alaska, the United States did not, despite appearances, take advantage of Canada. Alaska's current borders were already set when the United States purchased Alaska from Russia in 1867.

Alaska's Southern and Eastern Borders

In 1821, the Russian-American Company, a fur trading corporation operating under a charter from the czar, laid claim to all the land along the Alaskan coast north of the 51st parallel. This translated into a Russian presence as far south as Vancouver, a port that the British considered vital to their trade. To counter the Russian claim, the United States and England set aside their bitterness from the Revolution and the War of 1812 and became joint-claimants of a huge area of land known then as the Oregon Country. They claimed the northern border of the Oregon Country to be the 55th parallel.

In 1824, Russia essentially accepted the American/British claim, signing conventions with both nations. The first of these agreements, which was with the United States, set the southern border of Russian Alaska at 54°40'—an adjustment that preserved all of Queen Charlotte

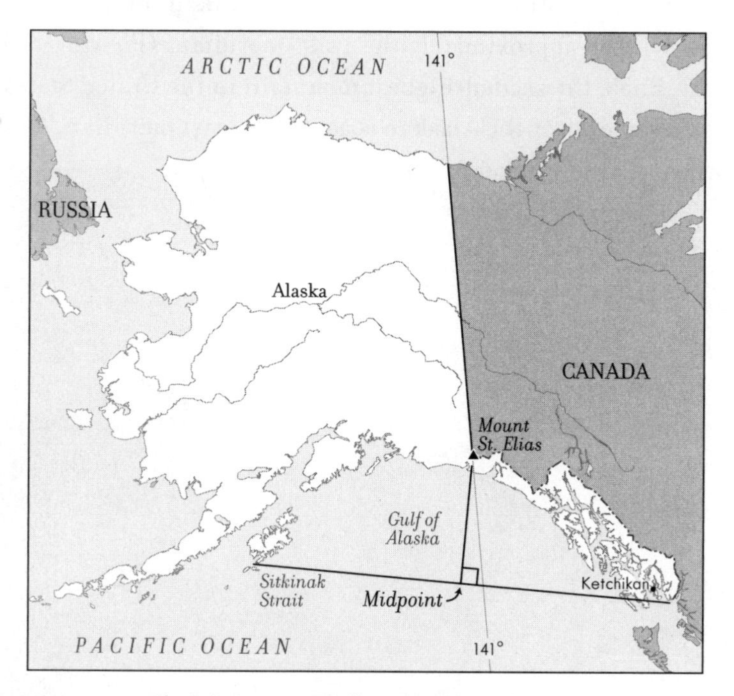

FIG. 20 Alaska's Eastern Border—the Reason for 141°

Sound and the port of Vancouver. (In the future, 54/40 would become an American rallying cry for war with England. For more on this event, go to OREGON.)

Russia's agreement with England over Alaska, which the two nations concluded after the American agreement, gave Russia access as far east as the 141st meridian but withdrew its southern boundary all the way back to the 60th parallel. With one important exception. The exception was that Russian Alaska kept the strip of land between the coast and the nearby mountains all the way down to the Ketchikan, at the 55th parallel. Why? And why the 141st meridian when 140 is such a nice round number?

The southern edge of Alaska forms a long arc, much of it protected by barrier islands. These barrier islands create safe harbors for ships and fishing boats, and those snug harbors are what the Russians most wanted. With this fact in mind, we can find the reason why Alaska's straight-line eastern border is located at 141° W longitude. If one bisects that segment of Alaska's coastal arc in which there are barrier islands (the segment from the Sitkinak Strait, on the west, to Ketchikan, on the east), the midpoint will be approximately the 141st meridian. (Figure 20) Plus, Mount St. Elias, the second-highest mountain in the United States and Canada, presides over the Alaskan coast at the 141st meridian. As such, it provides an ideal marker for a boundary.

ARIZONA

How come Arizona has that angle at the bottom aiming up to California? And how come that angle stops just short of California, ending in Mexico instead? Why are its straight-line borders located where they are?

The United States acquired the land that is today Arizona as part of its victory in the Mexican War, 1846–1848. (See Figure 28, in CALIFORNIA.) Initially, Arizona was part of the New Mexico Territory, the borders of which were set in 1850. (See Figure 119, in NEW MEXICO.)

Arizona's Southern Border

The boundaries of the New Mexico Territory changed on several occasions but only one of the changes affected what would become Arizona. In the early 1850s, many industrialists and investors sought to build a second transcontinental railroad across the southern tier of the country. In the southwest, however, only the Mesilla Valley provided a feasible pass through the mountains . . . and the Mesilla Valley was in Mexico. To make matters worse, Mexico and the United States were embroiled in a border dispute resulting from an ambiguity in the treaty ending the Mexican War.

Among those urging the construction of a second transcontinental railroad was James Gadsden, president of the South Carolina Railroad Company. In 1853, President Franklin Pierce appointed Gadsden United States Minister to Mexico. Gadsden's sole task was to solve the border dispute and the railroad dilemma in a single diplomatic stroke: buy the land. Gadsden did and the United States acquired what has since been known as the Gadsden Purchase. (Figure 21)

The Gadsden Purchase accounts for the angle in Arizona's southern border that heads northwestward toward California. This border appears as if it misses California, ending 20 miles to the south. Those 20 miles, which are specified in the treaty negotiated by James Gadsden, are the result of avoiding the Gila Mountains and of providing a buffer of

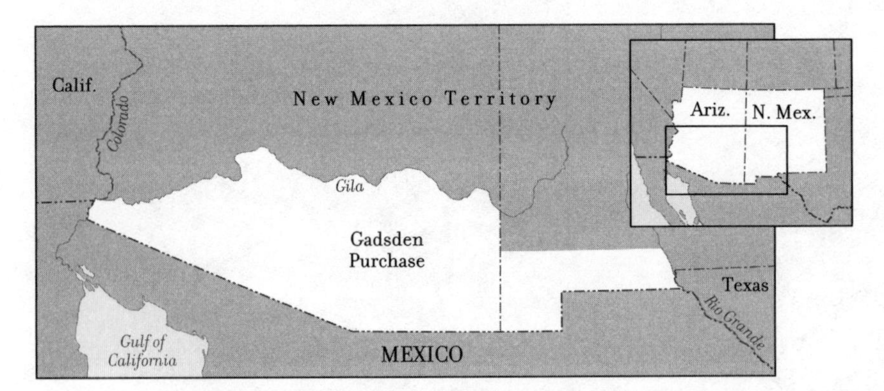

FIG. 21 The Gadsden Purchase—1853

land around the town of Yuma, California, where the Gila River joins the Colorado River. This juncture had great importance for commerce in the years when the Colorado was navigable from Utah to the Gulf of California, an era that ended with the construction of the Hoover Dam (completed in 1936).

Arizona's Eastern Border

Arizona's eastern border, dividing it from New Mexico, was created in 1863. But why was the territory divided vertically? And why is the vertical line where it is? As it happens, the *first* territory of Arizona was divided horizontally. In 1862, during the Civil War, an unofficial gathering of citizens in the New Mexico Territory voted on a declaration stating that the southern half of the territory, up to the 34th parallel, was joining the Confederacy. (Figure 22) At the time, the population of the southern half of the territory—the route of the Southern Pacific and the region where gold had been discovered—was predominantly Anglo, as opposed to Hispanic. A Confederate regiment soon arrived to defend the territory, followed by Union forces that quickly put an end to Arizona's secession.

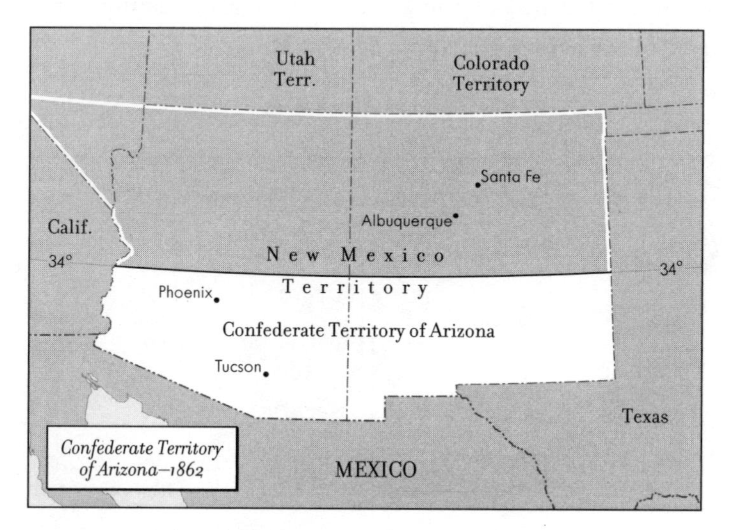

FIG. 22 Arizona's Shape as a Confederate Territory

When Congress created Arizona in 1863, it did so by dividing the New Mexico Territory *vertically* instead of horizontally. The vertical border demonstrated the quest by Congress to create equal states (a telling contrast to the horizontal border proposed by the territory's pro-slavery Confederates). The vertical border gave both Arizona and New Mexico access to the Gadsden Purchase with its valuable rail connections. In addition, the location of Arizona's eastern border divides the overall territory nearly equally between the two future states.

Arizona's Western Border

One year after the Civil War, steamboat navigation was opened on the Colorado River from the Gulf of California as far north as the town of Callville, Arizona. (Today, the only way to visit Callville is with a snorkel, since it's now at the bottom of Lake Mead, which was created by the Hoover Dam in 1936.) Immediately after steamboats came to Callville, Congress lopped it off Arizona and gave it to Nevada. In fact, Congress lopped off 18,000 square miles of Arizona—this being the triangle in Arizona's northwest corner—and added it to the southern end of Nevada. (Figure 23) Perhaps it was payback for the actions of Arizonans during the Civil War, but the land transfer also resulted in greater equality of size between Arizona and New Mexico.

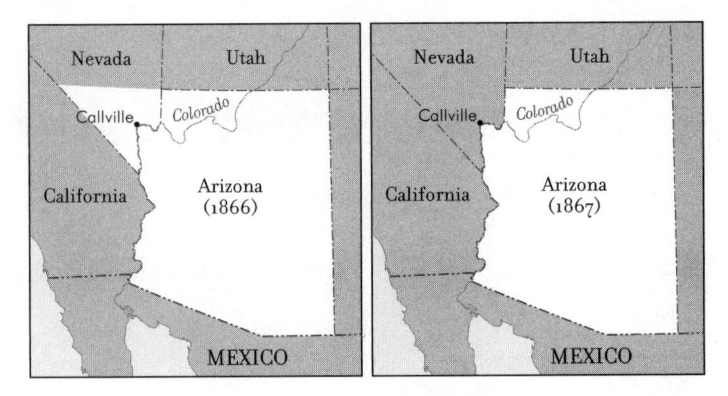

FIG. 23 Change in Arizona's Western Border

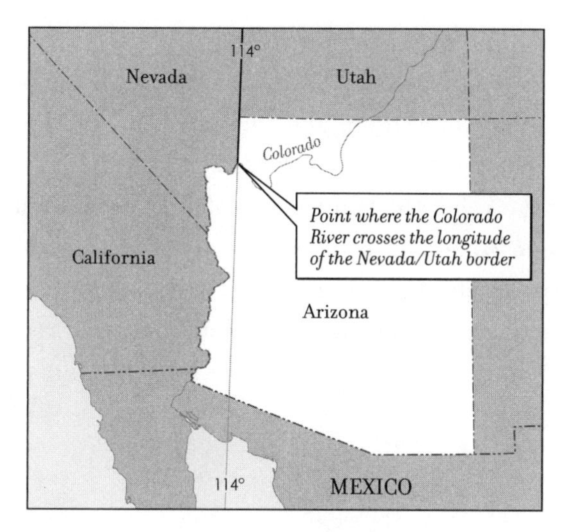

FIG. 24 Arizona's Straight-Line Northwest Corner

Arizona's western border was now the Colorado River to the point where it crossed the same longitude as the line between Nevada and Utah. (Figure 24) From there the border followed that line due north to its northern border, the 37th parallel.

Arizona's Northern Border

Why is the 37th parallel the northern border of Arizona? Arizona inherited this boundary from its days as part of the New Mexico Territory, but why was this the northern border of the New Mexico Territory?

The 37th parallel first emerged as a boundary in the Compromise of 1850, legislation that established the boundary between Texas and the New Mexico Territory. By establishing the northern border of the New Mexico Territory at 37°, Congress could then create a tier of Rocky Mountain states, each having four degrees of height, between the New Mexico Territory and the Canadian border. And indeed, Colorado, Wyoming, and Montana are all four degrees high.

Was this column of states, all with four degrees of height, really what Congress had in mind? Consider that nine years later, Colorado applied for statehood and proposed borders that would have made it somewhat larger than it is today. Because of its strong bargaining position (gold had just been discovered there and the Civil War was about to erupt), Colorado succeeded in obtaining borders that made it a bit larger than its neighboring states, but the two borders that Congress did adjust were Colorado's proposed northern and southern borders. It reduced them to 41° and 37°, precisely the latitudes required to maintain the equally spaced states envisioned when Congress established 37° as the northern border of the New Mexico Territory and the future state of Arizona.

ARKANSAS

Why are there notches in the northeast and southwest corners of Arkansas? And why is its western border a straight line that's bent? Since Arkansas' northern border (except for that notch) lines up with the Tennessee/Kentucky border, how come its southern border doesn't line up with anything?

The land that today comprises Arkansas was obtained by the United States in the 1803 Louisiana Purchase. This purchase created Arkansas' eastern border—the Mississippi River—since the Mississippi was the only clearly defined border of the Louisiana Purchase, other than the Gulf of Mexico.

Arkansas' Southern Border

The southern border of Arkansas was created one year later when President Thomas Jefferson proposed dividing the Louisiana Purchase along the 33rd parallel. The land to the south of the 33rd parallel was deemed the Orleans Territory (soon to become the state of Louisiana) and everything to the north was initially deemed the Louisiana Territory.

But why did Jefferson propose dividing the land at this location? With the acquisition of the Louisiana Purchase, President Jefferson had also acquired a new population of citizens—French-speaking citizens, for the most part. Jefferson was aware of the need to earn the trust and loyalty of this population. Toward that end, he and the Congress created a state for these new Americans that was small enough to govern effectively yet large enough to include all of the existing French settlements. At that time, the 33rd parallel represented the extent of those settlements in the lower region of the Louisiana Purchase. (For more details, go to LOUISIANA.)

Arkansas' Northern Border

In 1819, the population that had long been clustered at the juncture of the Mississippi and Missouri rivers (the city of St. Louis) was rapidly expanding. Congress responded by creating the territory of Missouri, which also resulted in the creation of the Arkansas Territory, lodged as it was between Missouri and Louisiana. Arkansas would, therefore, inherit as its northern border whatever southern border Missouri acquired. That border, as it turned out, immediately became historic.

On the face of it, the line used to divide Missouri and Arkansas was simply an extension of the line dividing its neighbors to the east: Kentucky and Tennessee. That line, in turn, was an extension of the line dividing Virginia and North Carolina. And that line, which originates at the Atlantic coast, midway between the Chesapeake Bay and Albemarle Sound, turns out to be 36°30' N latitude. But upon becoming the southern border of Missouri (and northern border of Arkansas) that line dovetailed with another American dividing line: slavery.

With the Louisiana Purchase, a question had arisen regarding slavery in the states to be created from this newly acquired land. Slavery had been prohibited in the states being created from the Northwest Territory (Ohio, Indiana, Illinois, Michigan, and Wisconsin). On the other hand, Congress allowed the state of Louisiana to continue the practice, since it had existed there prior to American acquisition. When Missouri sought statehood, being the first Louisiana Purchase territory to do so, it sparked a bitter dispute in Congress that resulted in the Missouri Compromise (1820). Under this agreement, only those states in the new territory whose northern borders were below 36°30' could sanction slavery—with the exception of Missouri.

The reason Congress chose 36°30' is that this preexisting line divided the United States (to the extent that it then existed) about as equally as one could gauge between present and future slave states and free states—given that the mountainous western end of the country was poorly suited to a slave economy, and given that some slave states already existed north of 36°30'. (Figure 25)

The northern border of Arkansas is interrupted by a notch in the northeast corner. The fact that this corner belongs to Missouri reflects attitudes regarding Arkansas when it was separated from Missouri. Plantation owners and other wealthy residents in what would have become Arkansas' northeast corner wished to remain with the more populated territory, since it was currently seeking statehood, and because of its powerful hub at St. Louis. So influential were these individuals that accommodations were made to extend the Missouri line below 36°30' by

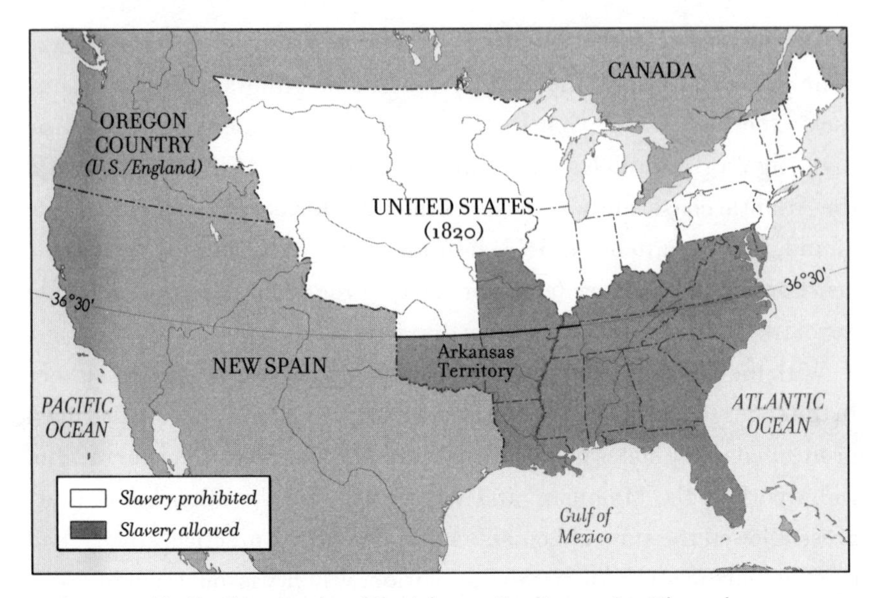

FIG. 25 The Northern Border of the Arkansas Territory—1820 Missouri Compromise

having it follow the St. Francis River south to the 36th parallel. (See Figure 103, in MISSOURI, where there are more details.)

Western Border of Arkansas

Also in 1819, the year Congress created Arkansas, the United States negotiated what was to become the notch in Arkansas' southwest corner. The Adams-Onis Treaty defined Spanish territory from the Louisiana Purchase territory. (Figure 26. See also DON'T SKIP THIS.) One segment of that division was a line due north from the point where the Sabine River crosses 32° N latitude. This turn in the border between the United States and Spanish-ruled Mexico accounts for the notch in the southwest corner of Arkansas. (For more details, go to LOUISIANA and TEXAS.)

But why leave a notch? When the Arkansas Territory was vertically

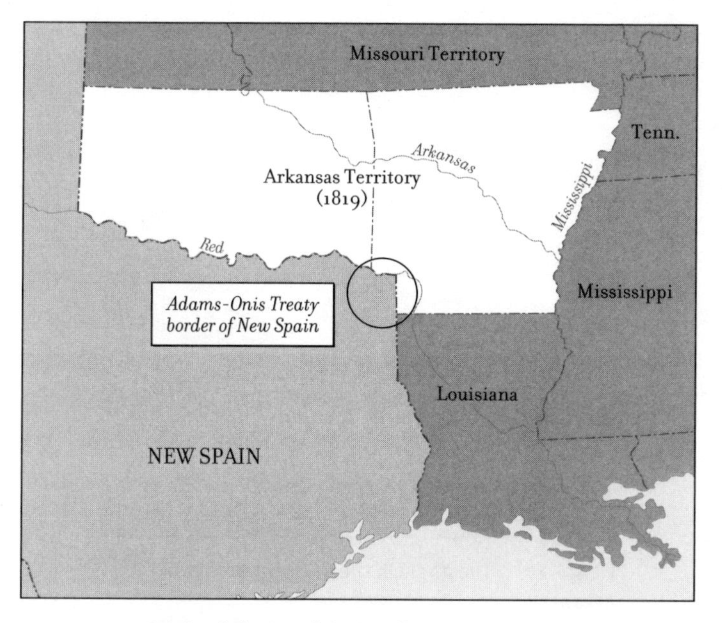

FIG. 26 Origin of the Notch in Southwest Corner of Arkansas

divided, creating Arkansas on the east and what would become Oklahoma on the west, why didn't Congress make the border between Arkansas and Oklahoma simply a continuation of the straight line coming up from the Sabine River in Texas (or a continuation of the straight line coming down from Missouri)? And why is the line that became the western border bent? Did someone goof?

Actually, someone did goof. And it was none other than Andrew Jackson. The line that serves as the western border of southern Missouri was indeed intended to continue as the western edge of Arkansas. But in negotiating with the Choctaws in 1820, Jackson inadvertently gave them far more of Arkansas than he realized. The Choctaws, after considerable urging (including a possibly involuntary "suicide" of one of their leaders), renegotiated the border in 1824, agreeing to relocate farther west, but not as far west as the western border of Missouri.

Under this later treaty, the eastern boundary of the Choctaw lands

FIG. 27 The Angle in the Western Border of Arkansas

began 100 paces west of the southwest corner of the main garrison at Fort Smith. After 100 paces, the lower half of the boundary extended due south to its intersection with the Red River. The upper half angled slightly on a straight line to the southwest corner of Missouri. (Figure 27) To this day, this line serves as the western border of Arkansas.

CALIFORNIA

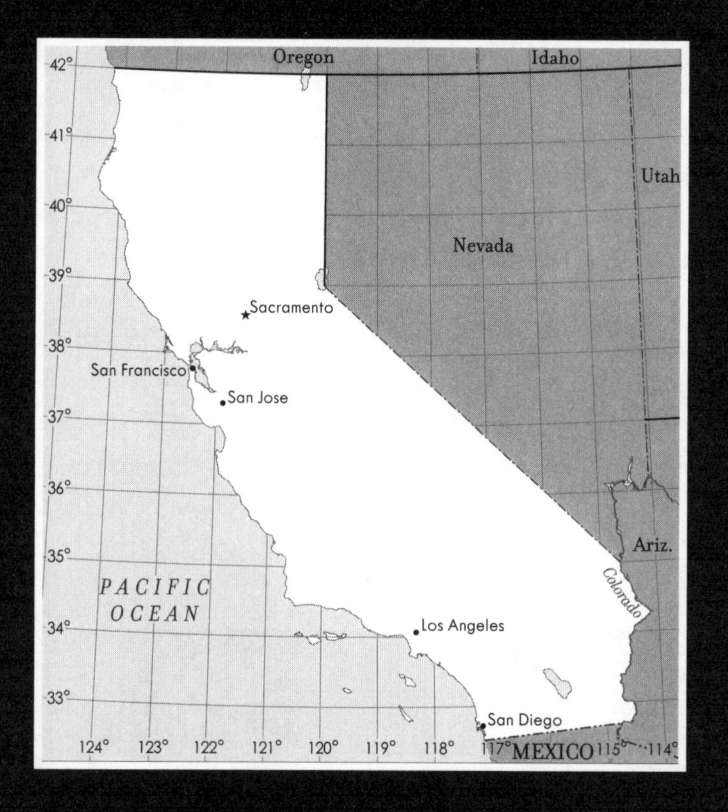

How come California is so big? And since it is so big, how come it doesn't include that long peninsula that continues from its southern end? Why are the straight lines of its northern and eastern borders located where they are? And why does its eastern border bend?

If Congress followed a policy that all states should be created equal, why did it create California? Answer: It didn't. California created itself. The land that became California came into the possession of the United

States in 1848 with the end of the Mexican War. (Figure 28) Before Congress could go through the process of dividing it into territories, a man named James Marshall spotted something shiny by the sawmill of his employer, John Sutter. It was gold.

Having been an American possession for barely a year, California was suddenly filled with a population, an economy, and a very high crime rate. So urgent was the need for government that Californians created their own state constitution and declared their own borders, skipping the fuss and bother of territorial status. Territories, after all, can have their borders altered; states (without their consent) cannot. Still, to become a state, Congress had to approve.

Congress said yes, despite the size of California and its wealth of natural resources. Indeed, if California were a nation today, it would have the fifth largest Gross National Product in the world. Why did Congress do it?

Many members of Congress were opposed to the extensive boundaries proposed by California. Nevertheless, many of those concerned by California's size voted in favor of its statehood. An echo of their reasoning

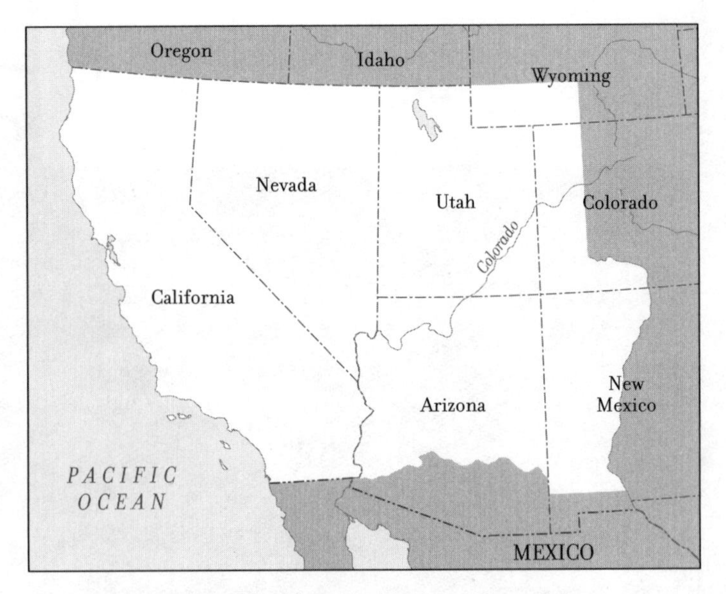

FIG. 28 Land Acquired from the Mexican War—1848

reverberates in that modern-day statistic just cited: *if it were a nation.* Today we look at the map of the contiguous forty-eight states and assume these states to be one nation, indivisible. In 1849, that assumption did not exist. Rather, there was considerable fear that the states might divide into separate nations. This was primarily a concern about the slave states. But bear in mind that our other oversized state, Texas, which had also only recently joined the Union, had previously been a nation for nearly ten years.

The same concern applied to California, as expressed aptly and emotionally on the floor of the Senate by William Seward (who, eighteen years later as Secretary of State, would purchase Alaska for the United States):

> [California] is practically further removed from us than England. We cannot reach her by railroad, nor by unbroken steam navigation. We can send no armies over the prairie, the mountain, and the desert. . . . Let her only seize our domains within her borders, and our commerce in her ports, and she will have at once revenues and credits adequate to all her necessities. Besides, are we so moderate, and has the world become so just, that we have no rivals and no enemies to lend their sympathies and aid to compass the dismemberment of our empire?

California violated the policy of equality among states because it could. The United States needed California more than California needed the United States. The size of its boundaries preserves these elements of mid-19th-century American life. The *location* of its boundaries preserves something more. Since they were dictated by California, they were located with the concerns of California in mind, not, as when Congress located borders, with the concerns of the region as a whole.

Why, for instance, did the state's founders include southern California? In those years, it contained far more desert than it does today, with the irrigation that has since been developed. The valuable harbor at San Diego certainly influenced the decision to extend the borders so far south. The importance of access to the ocean is also revealed by California's southeast border, the Colorado River. At the time that California established this boundary, the lower end of the Colorado River was navigable to the Gulf of California. Access to the Colorado River meant access to the sea.

California's Eastern Border

All of California's remaining borders are straight lines, raising the question, why there? The most striking of these straight lines is the long eastern border of California, a line that heads due south, then angles southeasterly until it reaches the Colorado River. The official reasoning for this line was that it paralleled, in a general way, the western border of the state at a distance of about 215 miles. True enough, but why 215? The answer is it encloses California's treasure.

California's eastern border is one of the few items from the Gold Rush that is still on the ground. Its existence is evidence of how important it was to California to possess all of the gold-bearing mountains in the region. Had Congress created the border, it might well have followed the crest of the mountains, as many of the eastern states have borders along the crest of the Appalachians. (Figure 29) Indeed, some years after

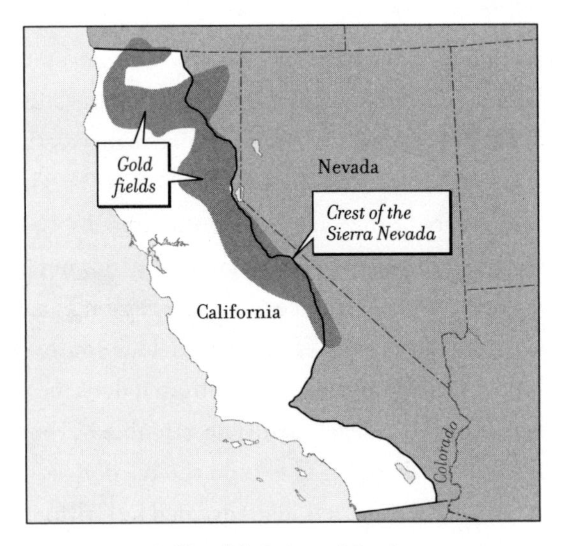

FIG. 29 California's Rejected Border

California had been admitted to the Union, Congress broached the idea of locating the California/Nevada border along the crest of the Sierra Nevada. California told Congress to forget it.

California's Northern Border

California's northern border is simply a straight line, going east and west, along the 42nd parallel. It is a segment of a boundary that was already in place before California existed. Under the Nootka Convention (1790) Spain and England agreed upon the 42nd parallel as the boundary between their Pacific coast claims. (To find out why this particular parallel was used, see DON'T SKIP THIS.)

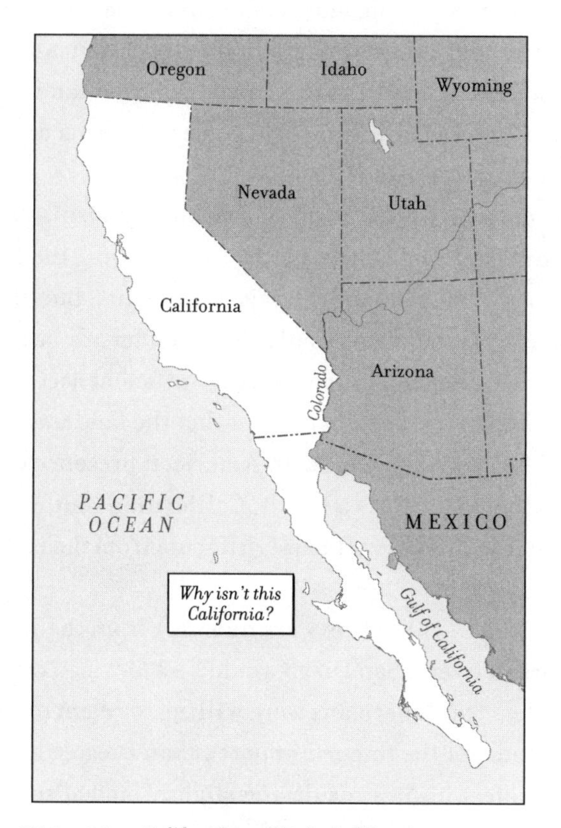

FIG. 30 California and Baja California

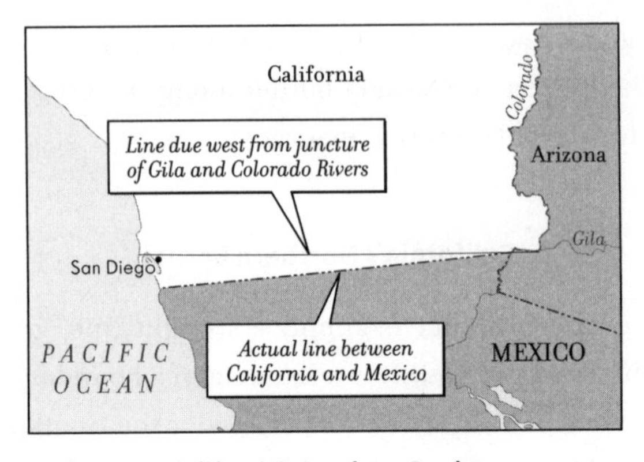

FIG. 31 California's Southern Border

California's Southern Border

At California's opposite end, how come this expansive state stops so abruptly at a straight line just below San Diego? Logically, shouldn't it continue on down that peninsula? (Figure 30) That land is even called Baja California—*Lower* California. And since California does stop there, why is the line slightly angled?

As it turns out, the United States *did* think California should extend down that peninsula. In negotiating the treaty ending the Mexican War, President James Polk demanded Baja California. But the Mexicans wouldn't budge. In fact, Mexico wouldn't surrender any land south of the Gila River. Mexico insisted on preserving sufficient access for its army to reach Baja California, despite the fact that the land was of little commercial use. The Mexicans feared an American presence on their west, in addition to their north. As a result, California's southern border is a slightly angled line that cuts off Baja California from the rest of the state. (Figure 31)

But why the angle? If the line went due west from the juncture of the Gila and Colorado rivers, San Diego would be a Mexican city located just below the border. The Americans were willing to relent on Baja California, provided they got the important port at San Diego, along with sufficient land to protect it. As a result, the slightly angled southern border was drawn.

COLORADO

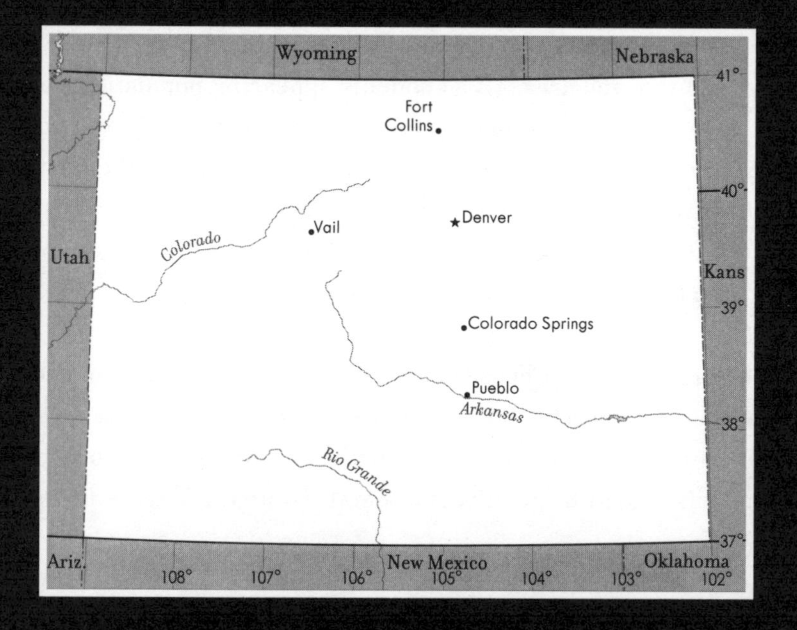

Why does Colorado have such boring borders? Why isn't its eastern border an extension of that nice long line dividing Nebraska from Wyoming and the Dakotas? Or if not that line, why isn't its eastern edge that nice long line dividing New Mexico from Texas and Oklahoma? And how come the northeast corner of Colorado takes a bite out of Nebraska, when Nebraska had been created first?

When Kansas and Nebraska were organized as territories in 1854, their boundaries extended to the crest of the Rockies, thereby encompassing much of what would eventually become Colorado. The rest of what would become Colorado was previously part of the Utah and New Mexico territories. (Figure 32) How sections of four separate territories come to be an entirely new territory can be answered in one word: gold.

In 1858, gold was discovered in what is now Colorado but what was then part of the Kansas Territory. Almost immediately, over 50,000 people flooded into the area. This suddenly appearing population needed closer access to, and responsiveness from, its territorial government. But the Kansas territorial government was absorbed in another matter—its debate over slavery (a debate in which views were expressed in ways that resulted in the nickname "Bleeding Kansas"). Consequently, delegates from the gold fields met in 1859 and created the "Territory of Jefferson."

The borders of the proposed Territory of Jefferson sought to encompass as much of the gold and silver waiting to be mined in that region as possible and to provide the territory with agricultural land, so as not to rely entirely on a finite quantity of mineral resources. To achieve this, the

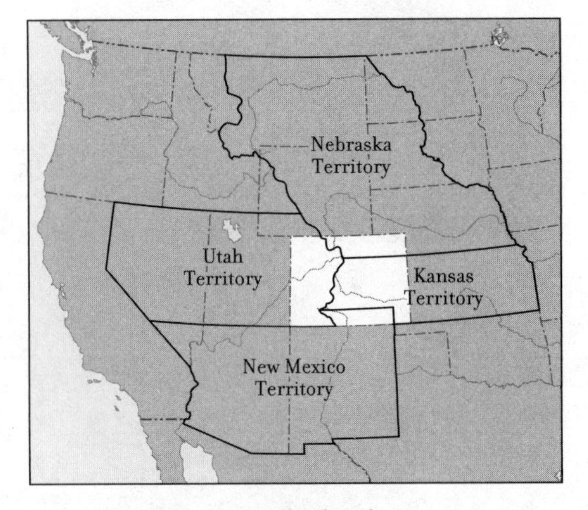

FIG. 32 The Origins of Colorado

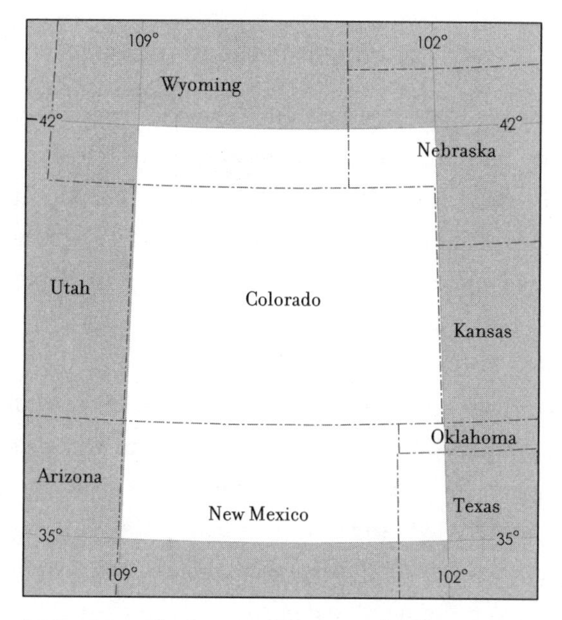

FIG. 33 The Proposed Territory of Jefferson

residents declared their southern border to be the 35th parallel and their northern border to be the 42nd parallel. On the west they claimed the land as far as the 109th meridian and on the east the 102nd. (Figure 33)

Congress created the official Territory of Colorado (as opposed to the unofficial Territory of Jefferson created by its residents) in 1861. In doing so, it accepted the territory's proposed eastern and western borders but not its proposed northern and southern borders. Why one and not the other? The answer gradually surfaced over the next fourteen years.

Colorado's Eastern and Western Borders

In the case of Colorado's eastern and western borders, the reason Congress accepted the territory's proposal became apparent with the later creation of the states of Washington (1889), North Dakota (1889), South Dakota (1889), and Wyoming (1890). Like Colorado—and the previously created state of Oregon (1859)—all of these western states have seven degrees of width. (See Figure 13, in DON'T SKIP THIS.)

Colorado's Northern and Southern Borders

Some of the reasons Congress rejected the territory's proposed southern border were immediately apparent. Located at 35°, it had been chosen by the territory because it would encompass more gold. (Figure 34) But it would have also encompassed America's recently acquired Hispanic-populated Santa Fe, along with (unconstitutionally) a corner of the State of Texas.

The proposed northern border of 42° corresponded to a preexisting border that dated back to a 1790 agreement between England and Spain known as the Nootka Convention. (Figure 35. For more on the Nootka Convention, see DON'T SKIP THIS.) The reason it was rejected would also become apparent over the next fourteen years.

Congress retracted the southern border of the proposed territory from 35° to 37°, and the northern border from 42° to 41°. These adjustments revealed Congress once again looking ahead, as it had with the

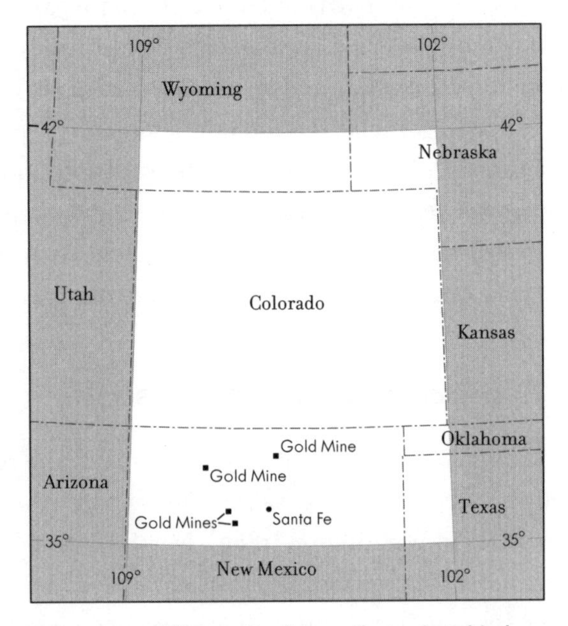

FIG. 34 Jefferson Territory—the Logic Behind the Southern Border

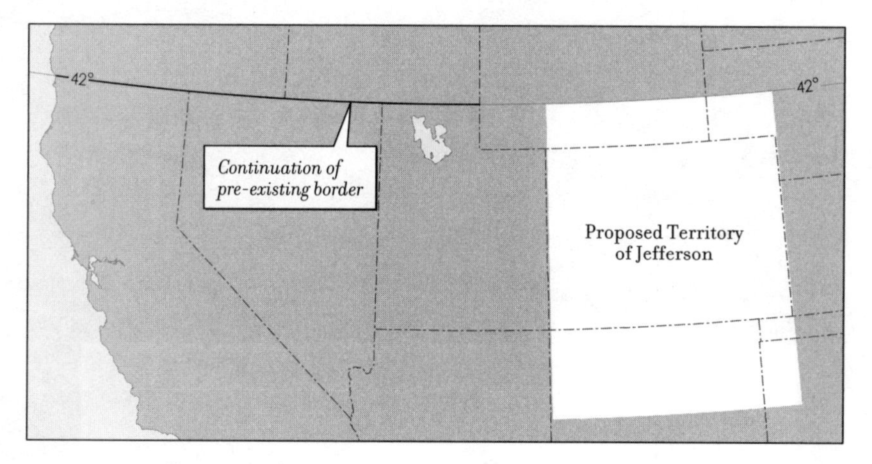

FIG. 35 Jefferson Territory—Logic Behind the Northern Border

eastern and western borders of Colorado. In fact, Colorado's southern border had already been envisioned in 1850, when Congress located the northern border of the New Mexico Territory at 37°. This location enabled a tier of three Rocky Mountain states to be created north of New Mexico, each with four degrees of height. In becoming states, Wyoming (1890) and Montana (1889)—both with four degrees of height—would join Colorado to fill the space between New Mexico and the Canadian border. (See Figure 12, in DON'T SKIP THIS.)

The northern and southern borders of Colorado are artifacts of something remarkable. Or perhaps they are artifacts of something we think is remarkable, but which goes on more often than we realize. They are artifacts of foresight and planning by our elected representatives.

CONNECTICUT

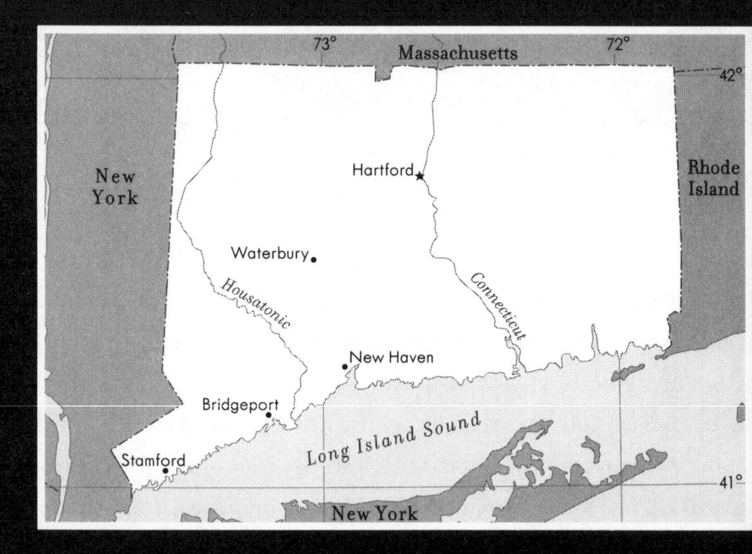

Why does Connecticut have that little panhandle hanging down on its west side? How did it get that notch in its northern border? How come the western borders of Connecticut and Massachusetts don't quite connect? And, why isn't Connecticut—and Rhode Island, for that matter—just part of Massachusetts?

The area that is today called Connecticut was previously pa Massachusetts—or, more specifically, it was part of the Plymouth C

Its existence as a separate entity evolved over time. Connecticut's evolution began in the 1630s when members of the Plymouth Colony established trading posts and small settlements along the rivers in the region. These were created at some risk, since the area was populated by the Pequot Indians. The conflicts between the colonists and the Pequot culminated in the Pequot War, which ended in 1637 with the defeat of the Pequot. Colonists from the Plymouth and Massachusetts Bay colonies poured into the newly won land. Their settlements began along Long Island Sound and spread up along the rivers that flow through the region into the Sound. In effect, they were creating the shape of Connecticut.

Connecticut's Northern Border

Within two years of the Pequot War, the residents of Hartford drew up the Fundamental Orders of Connecticut, and in 1642, the de facto existence of Connecticut as a separate colony was accepted by the colonists in Massachusetts, which dispatched surveyors to locate the border between Connecticut and Massachusetts. From the point of view of the colonists in Massachusetts, their fellow Puritans who created Connecticut were furthering their mission to create a New Jerusalem. From the point of view of the colonists in Connecticut, Massachusetts drew the border eight miles too far south.

But what was the southern border of Massachusetts, which had recently been formed by the merger of the Plymouth and Massachusetts Bay colonies? The Plymouth Colony's charter put its southern border at 40° N latitude. That turned out to be the latitude of Philadelphia. The Massachusetts Bay Colony's charter described its southern border as a line due west from a point "three English miles to the southward of the southernmost part of the said Bay called Massachusetts." Massachusetts argued for a line due west from a point 3 miles south of the town of Plymouth. Connecticut opted for a line due west from a point 3 miles to the south of the southernmost reach of the Charles River. (Figure 36)

Ultimately, Massachusetts accepted Connecticut's approach but not

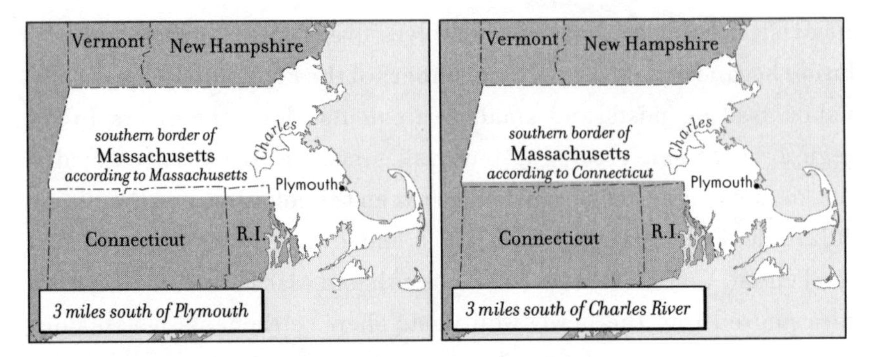

FIG. 36 The Disputed Connecticut/Massachusetts Border

the river, since the Charles is not the southernmost waterway leading into Massachusetts Bay. The Bay's southernmost tributary is the Neponset River. Hence Connecticut's northern border is formed by a line due west from a point 3 miles south of the southernmost point of the Neponset River. (Figure 37)

This negotiation was prolonged by the fact that the towns of Enfield, Somers, Suffield, and Woodstock, which existed well before Connecticut, were located in the disputed zone. (Figure 38)

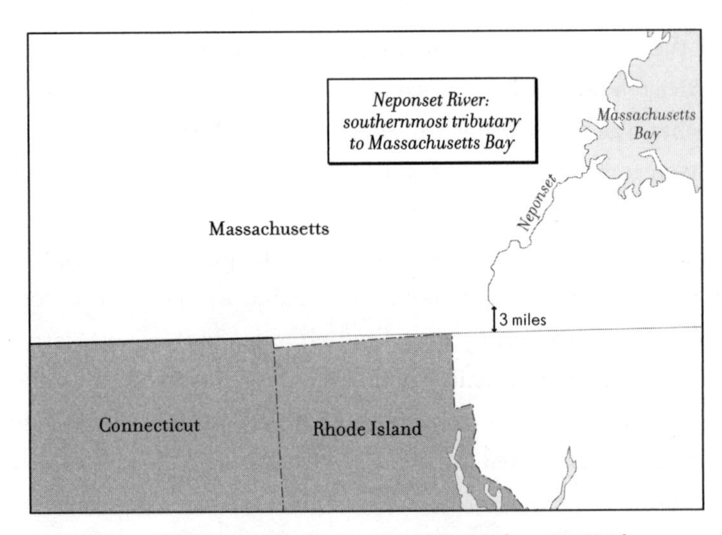

FIG. 37 The Formula for Connecticut/Massachusetts Border

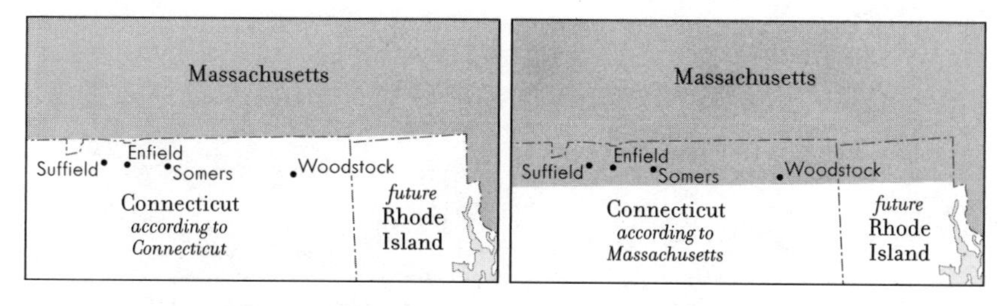

FIG. 38 Massachusetts Towns Inside Connecticut Border

Finally, in 1804, the two sides agreed that, as compensation to Massachusetts for losing these towns, Connecticut would partition Congamond Lakes, farther west. This is why there is a notch in the northern border of Connecticut. (See Figure 91, in MASSACHUSETTS.)

Connecticut's northern border also contains a very slight dip, east of the notch. This dip reflects a final concession to Massachusetts. Connecticut agreed to let the boundary follow the crest of the hills at the point where the Connecticut River crosses the border. Today the gesture may seem minor, but in an era when the Connecticut River and its riverbanks were vital to the region's prosperity, the more geographically natural boundary was significant.

Connecticut's Western Border

During the years that Connecticut was disputing its border with Massachusetts, it was also arguing with New York. This conflict emanated from a charter bestowed upon Connecticut in 1662. King Charles II granted Connecticut all the land bounded on the east by Narragansett Bay, on the north by the Massachusetts Colony, on the south by Long Island Sound, and on the west by, as had become the tradition, the Pacific Ocean. (Figure 39)

It didn't matter that this boundary overlapped Dutch claims, since England and Holland repeatedly went to war over their American territories. But when England ousted Holland for good in 1674, turning its land into Delaware, New Jersey, and New York, Connecticut's Pacific

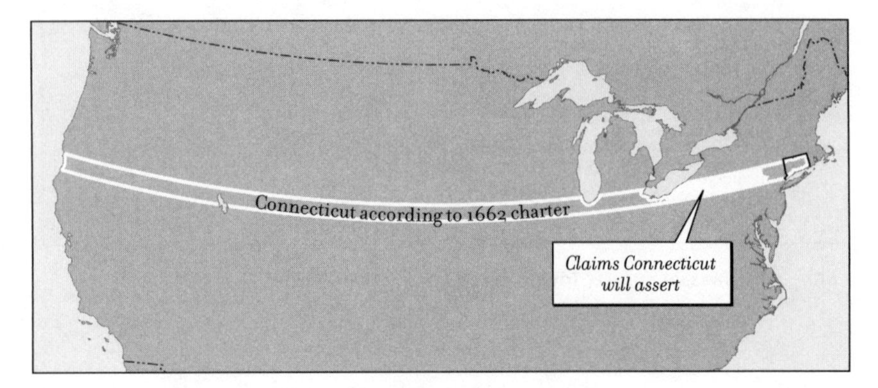

FIG. 39 Connecticut According to 1662 Charter

coast border ruffled the feathers of New York. (See Figure 114, in NEW JERSEY.)

The two colonies commissioned a boundary survey in 1683. The problem was providing New York with the agreed upon 20-mile buffer east of the Hudson while at the same time preserving for Connecticut its towns of Greenwich and Stamford. The solution turned out to be a panhandle. (Figure 40)

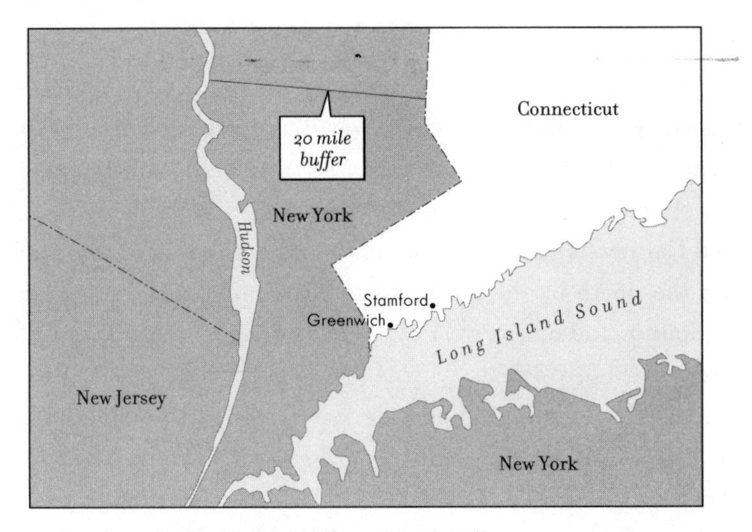

FIG. 40 Connecticut's Southwest Panhandle

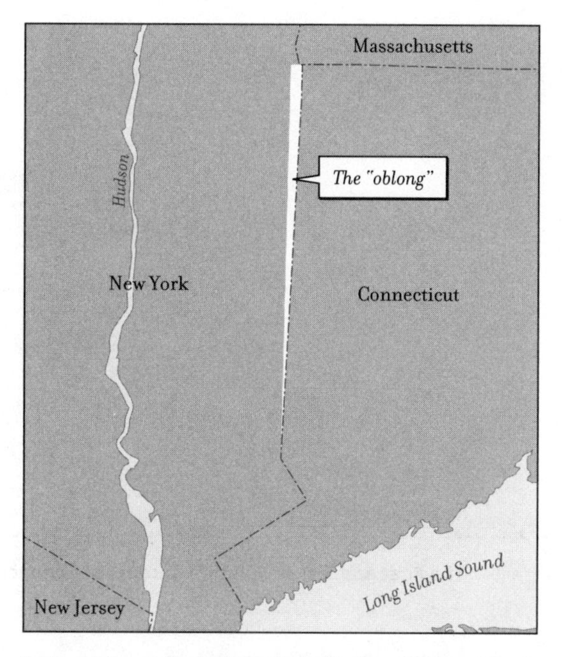

FIG. 41 Adjusted Western Border of Connecticut

To compensate New York for the land claims it released in creating the panhandle, Connecticut gave to New York a strip of land along its western border. This land is known as "The Oblong." (Figure 41) It extends from Ridgefield, Connecticut, to the Massachusetts line. It is this strip of land that accounts for the fact that Connecticut's western border is not quite aligned with that of Massachusetts.

Connecticut's Eastern Border

On the east, the border that King Charles II had stipulated in his 1662 Connecticut charter, he un-stipulated in his 1663 Rhode Island charter. To create the colony of Rhode Island, Charles II fixed as its boundary with Connecticut the Pawcatuck River, some 20 miles west of Narragansett Bay, which had been Connecticut's eastern edge. The king now said that the boundary line was to follow the Pawcatuck to its source, then continue due north to the Massachusetts line. (Figure 42)

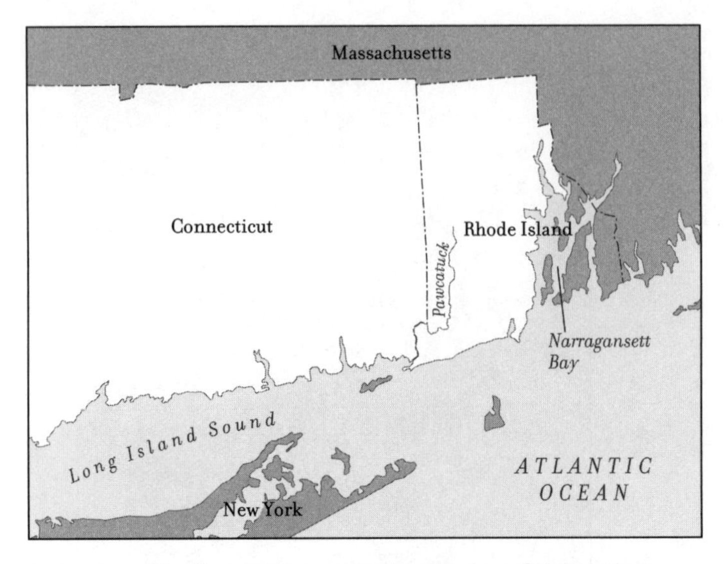

FIG. 42 **The First and Second Eastern Borders of Connecticut**

Connecticut's Western Lands

Boxed in by the powerhouse colonies of New York and Massachusetts, Connecticut turned its attention in the mid-1700s to its western land. The colony backed a corporation called the Susquehanna Company, which commenced to settle Connecticut's land west of New York. Today we call this land Pennsylvania and even then Pennsylvania called it Pennsylvania. Connecticut entered into yet another dispute. In fact, it entered into a war, known as the Pennamite War, in which its settlers and Pennsylvania's settlers opened fire upon one another in a number of skirmishes.

Less controversial were Connecticut's efforts to develop its lands further west, in what would eventually become Ohio. Since neither of these regions is considered a part of Connecticut today, what happened?

With the onset of the Revolution, the one-time colonies, now states, began to think *nationally* in addition to locally. The federal government urged those states with extensive colonial land claims to donate those lands to the United States so that more states could be created, more equal in size, and so that the government could use the land to raise

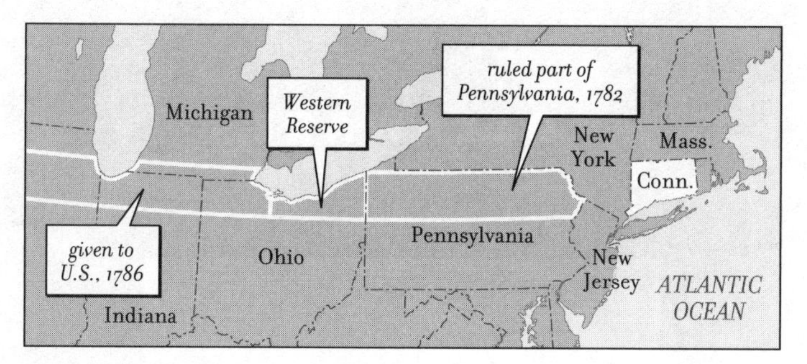

FIG. 43 Connecticut's Western Land Claims

funds to help retire the enormous debt it had acquired during the course of the war. Connecticut relinquished most of its claims but insisted on retaining its claims to the lands that its Susquehanna Company had purchased from the Indians and begun selling to investors, much of which was also claimed by Pennsylvania. The dispute ended up in a special court of arbitration that ruled in favor of Pennsylvania. (Figure 43) Connecticut did not immediately accept this decision and the Pennamite War resumed. But the movement toward nationhood was already in motion, and Connecticut ultimately accepted the decision in favor of Pennsylvania.

Connecticut continued, however, to retain its Western Reserve, land in what is now Ohio. But the investments were not proving profitable, and in 1800, Connecticut released this land, too. Still, a shadow of Connecticut's presence remains in Ohio, in the Cleveland-based university named Case Western Reserve.

Why is there a semicircle at the top of Delaware? And why does the straight line running south from that circle actually start just a little bit west of it? Why isn't Delaware simply part of Maryland? And since there is a Delaware, why aren't its borders the natural ones—the ocean and Delaware Bay on the east, the Chesapeake Bay on the west, all the way down that long strip of land?

Delaware isn't part of Maryland because at the time King Charles I granted Maryland a charter, in 1631, the area that would later be called Delaware already existed. And it was Dutch. Holland laid claim to the entire region between the Delaware River and Bay on the west and the Connecticut River on the east. (See Figure 114, in NEW JERSEY.) England, on the other hand, laid claim to all of North America, which led to conflicts that, in 1674, were resolved when England ousted the Dutch from the last of their North American land.

With Dutch authority gone, William Penn appealed to the crown to attach the territory that later became Delaware to Pennsylvania. Unlike all the other colonies, Pennsylvania had no window on the Atlantic, leaving it vulnerable to whoever controlled Delaware Bay. (Figure 44)

Maryland, however, believed it owned this area, since Maryland's royal charter defined its boundaries as including all the land on the Atlantic coast from a point due east of Watkin's Point up to the 40th parallel. (Figure 45) The fact that the 40th parallel ran right through Philadelphia would cause conflicts with Pennsylvania, but Delaware, from Maryland's point of view, was clearly within its domain.

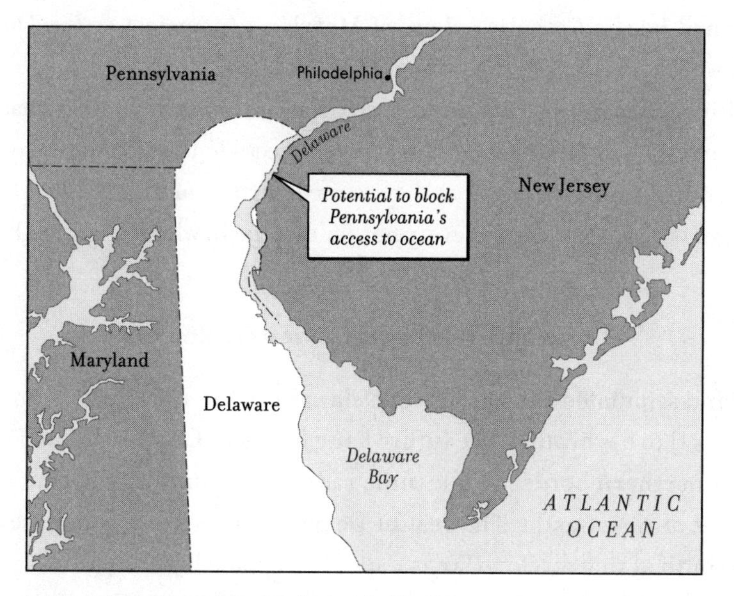

FIG. 44 Pennsylvania's Access to the Ocean

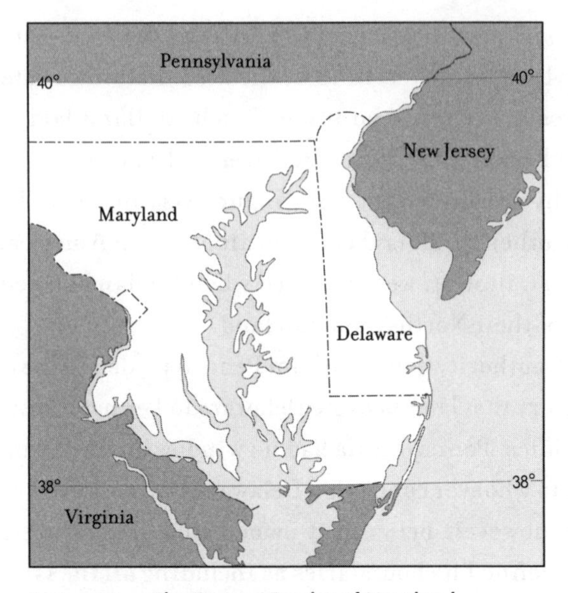

FIG. 45 The Eastern Border of Maryland According to its 1632 Charter

The residents of Delaware were opposed to being swallowed by either colony. As members of the Dutch Reformed Church, they were loath to be governed by the Catholic colony of Maryland (papists!) or the Quaker colony of Pennsylvania (heathens!).

Ultimately, the king decided to lease Delaware to Pennsylvania. That way, William Penn got what he wanted: assured access to the Delaware Bay. The residents of Delaware got what they wanted: virtual autonomy. And Maryland got nothing (as it did in every border dispute in which it engaged).

Delaware's Northern and Western Borders

England stipulated the borders of Delaware in the 1682 deed from King Charles II to his brother, the future King James II. Charles defined Delaware's northern border as a 12-mile radius surrounding the Dutch settlement at New Castle. The rest of Delaware, he stated, comprised the land south of that circle as far as Cape Henlopen. (Figure 46)

Maryland lodged a formal protest. In response, the King's Board of

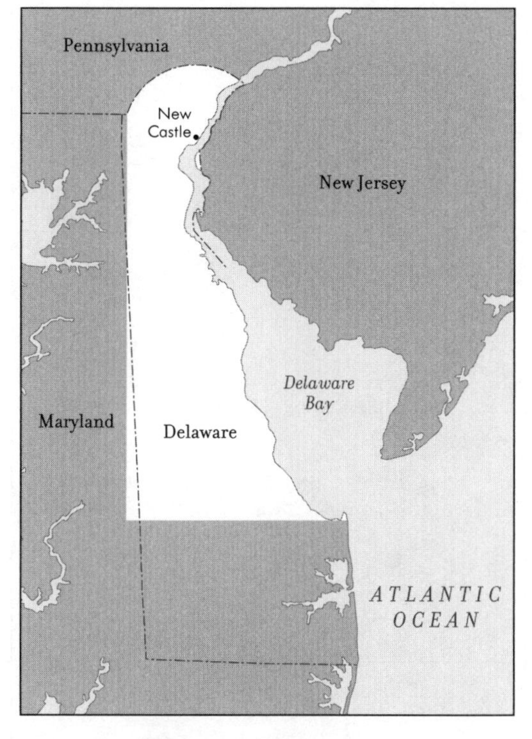

FIG. 46 Delaware—1682

Trade and Foreign Plantations ruled in 1685 that the charter creating Maryland was intended to include only land *uninhabited by Christians* at the time the charter was issued (which, indeed, the charter does say). The Board then sought to clarify the border, but the boundaries turned out to be anything but clear—or even possible:

> . . . for avoiding further differences, the tract of land lying between the River and Bay of Delaware and the Eastern Sea on the one side and Chesapeake Bay on the other, be divided into two equal parts by a line from the latitude of Cape Henlopen to the 40th degree of northern latitude.

As those involved were soon to discover, the point where the Delaware River crosses 40° N latitude is the Philadelphia docks. They would also

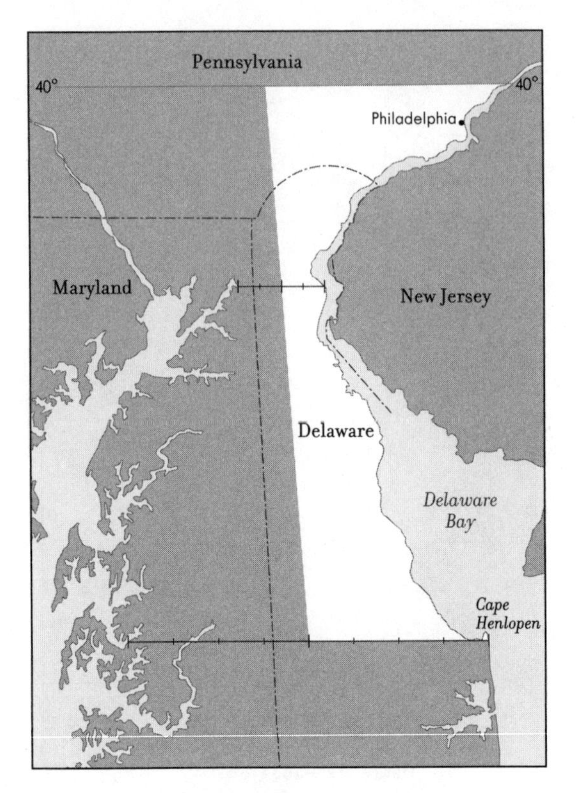

FIG. 47 Delaware—1685

discover that a line bisecting the land between the east coast and the Chesapeake from the latitude of Cape Henlopen to the 40th parallel slices through the 12-mile radius around New Castle. (Figure 47)

Delaware's western border was adjusted to be tangent to its 12-mile radius. But this line resulted in a sizable wedge among the neighboring borders, and this pocket of uncertain jurisdiction invited all sorts of individuals engaged in behaviors that jurisdictions tend to "jurisdict." (Figure 48)

To eliminate this problem, the center point of the arc was relocated in 1750 from the steeple of the old Dutch church to the courthouse and the smaller "wedge" that resulted was given to Delaware. This adjustment accounts for the fact that Delaware's northwestern border appears as if it just missed the state's northern arc.

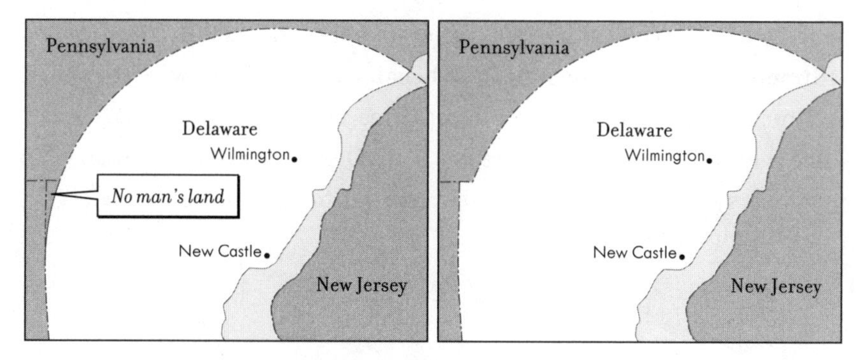

FIG. 48 **Adjustment to Eliminate "No Man's Land"—1750**

Delaware's Southern Border

Cape Henlopen, the specified southern extent of Delaware, is not the southern extent of Delaware. How did Delaware get the extra land? Delaware got it because Lord Baltimore, the colonial governor of Maryland, was working from an incorrect map. On his map, Cape Henlopen was located where Fenwick Island is today. This error accounts for the location of Delaware's southern border at Fenwick Island rather than Cape Henlopen, nearly 25 miles to the north. Despite Maryland's efforts to have it rectified, Delaware's southern border has remained at Fenwick Island to this day. (See Figure 87 and more details in MARYLAND.)

Delaware's Eastern Border

Delaware's eastern border was defined in 1674 when its neighbor to the east, New Jersey, was created. The boundary between Delaware and New Jersey was defined by its proprietor, the Duke of York, as the eastern shore of the Delaware River and Bay. Thus the river and the bay were part of Delaware and, to the north, Pennsylvania.

Today, however, two small patches on the Jersey side of the river are actually Delaware! (See Figure 2, in INTRODUCTION) Both are the result of river dredging to maintain sufficient depth for shipping. When dredging commenced in the early 20th century, the sediment was deposited in the river alongside the Jersey shore. Eventually, the dumped sediments rose above the water and became part of the New Jersey shore. Technically, however, the two dump sites remained part of Delaware, since they were created from areas within Delaware's river boundary. No controversy ensued, most likely because neither area is habitable.

DISTRICT OF COLUMBIA

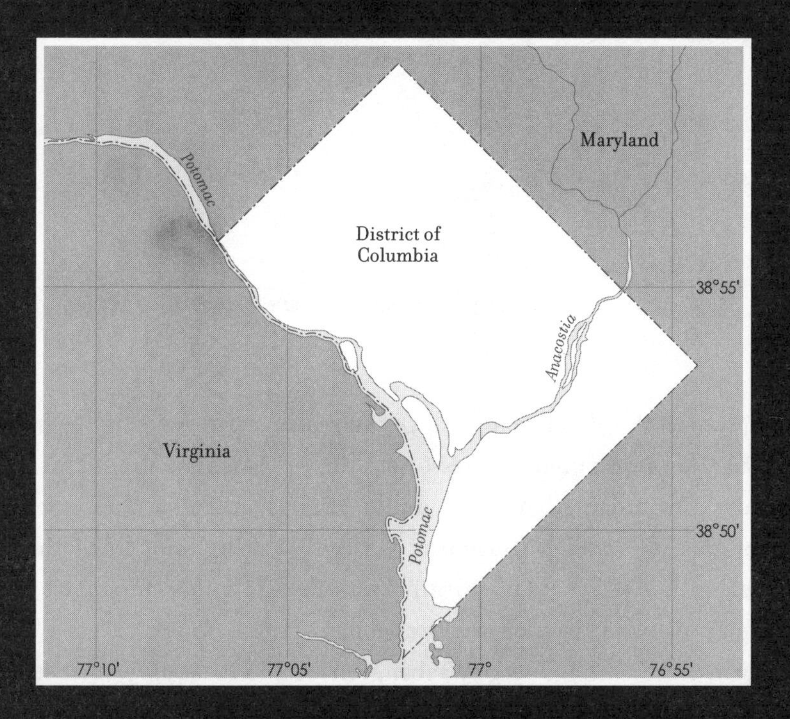

How come Washington, D.C., is only a partial square? And why are its straight lines located where they are? Why not farther up the Potomac or farther down? Or closer to the Potomac or farther away?

In 1790, Congress authorized President George Washington to locate a district of 100 square miles anywhere along the Potomac River between the mouth of the "Eastern Branch" (known now as the Anacostia River) and Conigogee Creek (known now as Conocheague Creek), near present-day

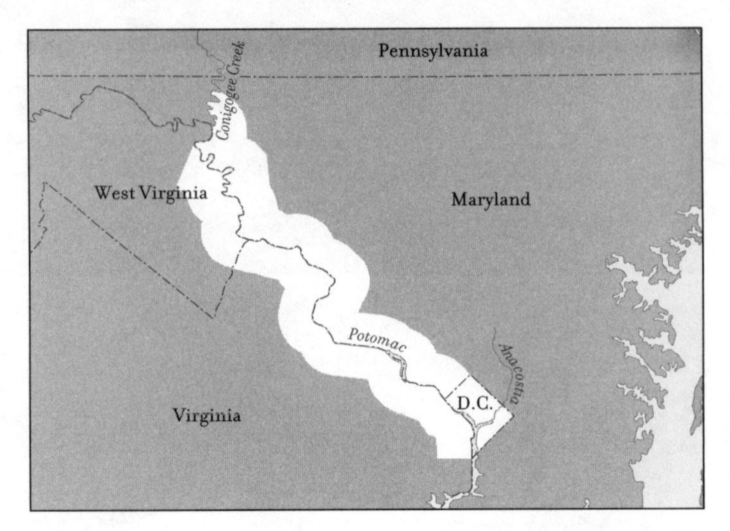

FIG. 49 The Area Along the Potomac Available for the Nation's Capital

Hagerstown, Maryland. (Figure 49) The reason for these parameters was that just beyond Conigogee Creek were the Appalachian Mountains, presenting a formidable barrier to diplomatic travel. Below the Anacostia, prosperous plantations (including George Washington's) lined the Potomac. Neither Maryland nor Virginia was likely to donate ten square miles of riverfront owned by such wealthy people.

Given these limits, George Washington found a 10-square-mile stretch with excellent potential, located between the Maryland community of Georgetown and the mouth of the Anacostia. From Georgetown on up, the river was not navigable. Unfortunately, below Georgetown, the river encountered the ocean's tidal flux. The daily high tides traveled up the Chesapeake and, in turn, up the Potomac, meeting the river's current and resulting in a backwater of swamps just below the rapids of the Potomac. This bug-ridden backwater became the nation's capital.

Positioning the District's Boundary

Even amid the swamps, President Washington envisioned a city whose boundaries could jumpstart its economy by incorporating two existing

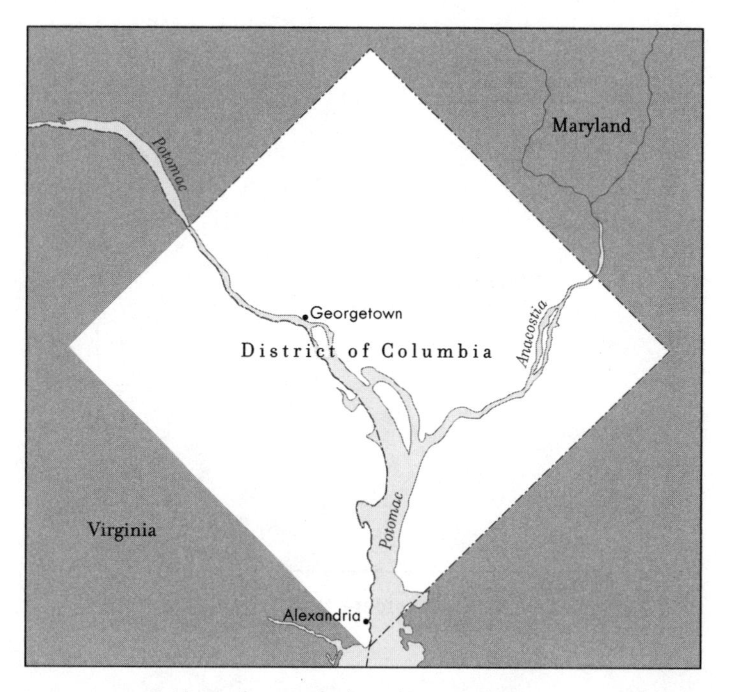

FIG. 50 **D.C.'s Border—Inclusion of Preexisting Ports**

ports: Alexandria, on the Virginia side, and Georgetown, on the Maryland side. (Figure 50) Given these population centers, Washington then oriented the 10-square-mile boundary in a way that maximized the potential to generate commerce.

The District's Northeast Border

Upriver from the point where it empties into the Potomac, the Anacostia River is navigable for about 5.5 miles. Washington located the northeast borderline as near as possible to the point where the Anacostia ceases to be navigable. (Figure 51) In this respect, the boundaries are an artifact of a time (short-lived to be sure) when Americans did not think of the federal government as a possible source of income for those gathered about it.

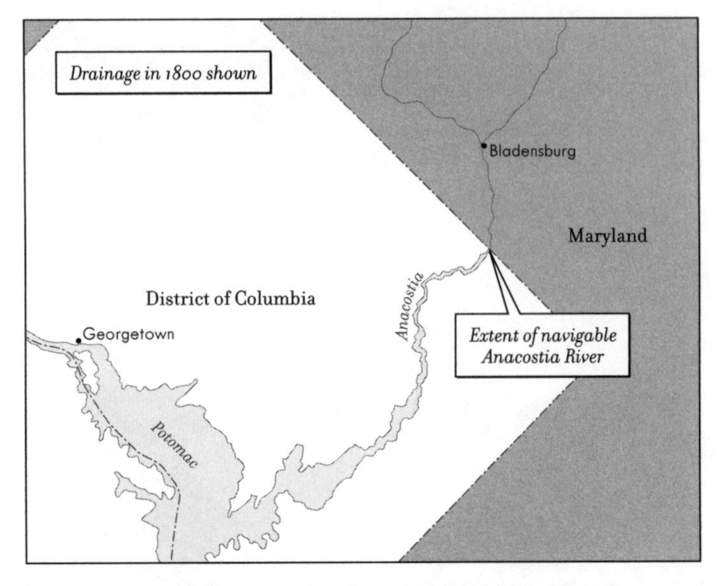

FIG. 51 D.C.'s Northeast Border—the Navigable Anacostia

The District's Northwest Border

The reason President Washington did not similarly locate the city's northwestern border at Georgetown, the farthest navigable point on the Potomac, had to do with the fact that he was working with a square. Had he fixed the northwest border at Georgetown, instead of 3 miles farther upstream, the southeast border would have been 3 miles below the Anacostia, which would have exceeded the available segment of the Potomac. (Figure 52) In addition, fixing the northwest border of the District at Georgetown would have resulted in the city being almost entirely on the Maryland side of the Potomac.

Even with the configuration he chose, the bulk of the District lay north of the Potomac. Why didn't Washington even things out by adjusting the square accordingly? He didn't because any adjustment to add land from Virginia would have sacrificed either some of the navigable Anacostia or the port at Alexandria.

The job of conducting the actual survey of the boundary went to Benjamin Banneker, a free African American. In contrast to the borderlines

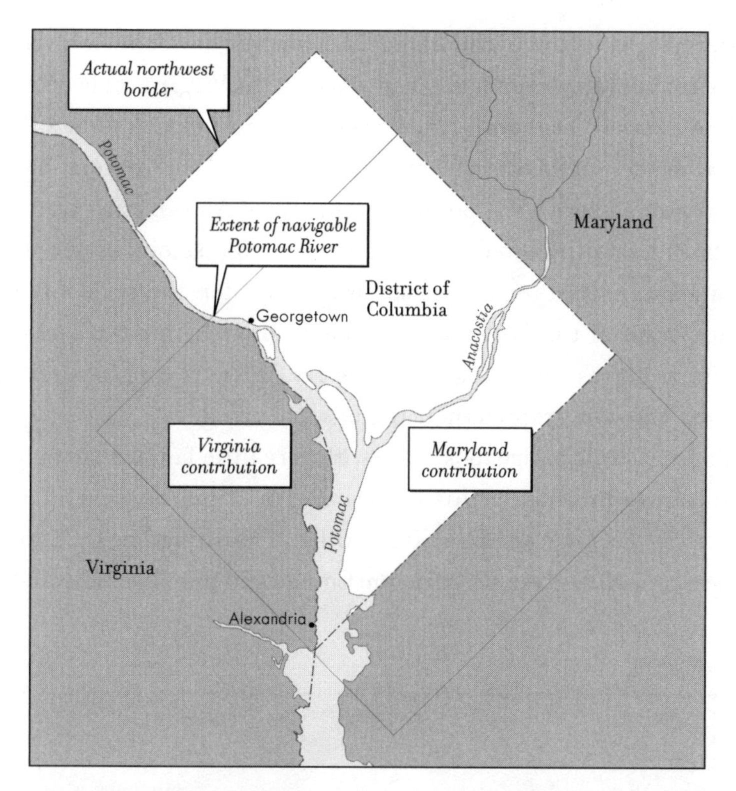

FIG. 52 D.C.'s Northwest Border—the Logic of Its Location

of Virginia/North Carolina, Connecticut/Massachusetts, Kentucky/Tennessee, and elsewhere, which later proved faulty, Banneker's boundary lines have stood the test of time. As such, the Washington, D.C., border is a very valuable artifact of the achievements of 18th-century African Americans in the face of overwhelming adversity.

The District's Southwest Border

Men like Benjamin Banneker were one of the reasons District residents south of the Potomac later sought to have their side of the city returned to Virginia. Free African Americans could live and work in the District of Columbia, but Virginia limited their stay to six months. Free African Americans thus participated in the commerce of the District. But during the first fifty-six years of the District's economic development, virtually none of that development had taken place on the Virginia side of the city. White workers in economically depressed Alexandria resented the additional burden of having to compete with free black workers. To make matters worse, from their point of view, one of Alexandria's traditional industries was eliminated in 1844, when Congress outlawed the slave trade in the District of Columbia. Two years later, the residents of the District south of the Potomac petitioned Congress for retrocession to Virginia. Congress, with bigger fish to fry (it was frying Mexico, at the moment), casually granted their request. (Figure 53)

This boundary change took place with very little fanfare, though there were those who proclaimed this act a crime. The crime scene is still on view today in the peculiar semi-square outline of Washington, D.C., and the outlines, across the river, of Arlington County and the city of Alexandria.

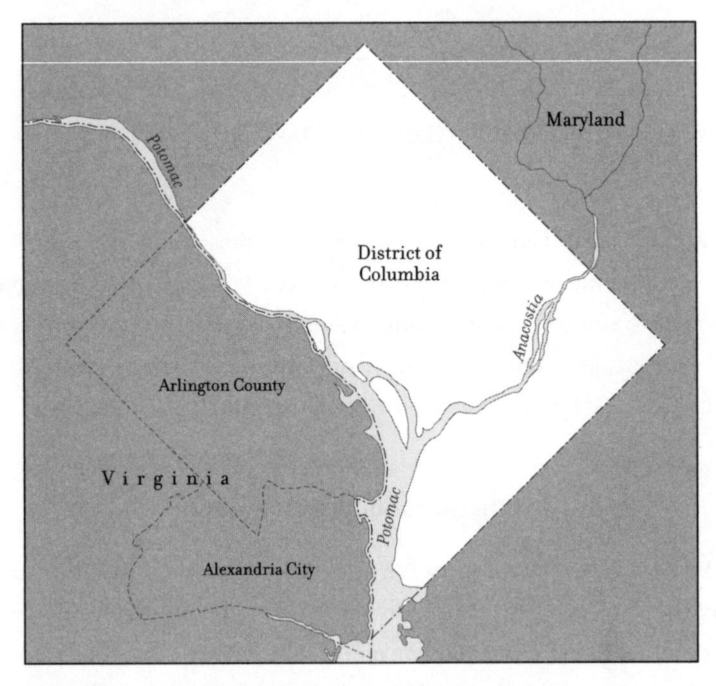

FIG. 53 D.C. After Retrocession to Virginia—1846

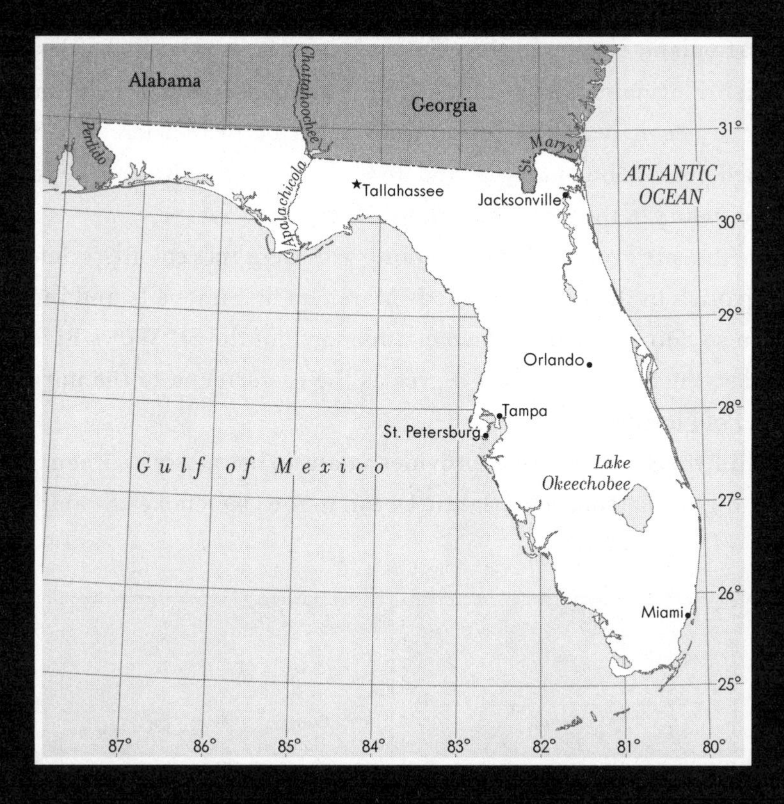

Aren't the reasons for Florida's borders pretty obvious? Still, how come two different straight lines define Florida's northern border? And why, at the eastern end of its northern border, does the boundary abandon its straight line and dip down, then jog up?

Florida was originally part of Spain's colonial territories in the New World, an empire that included all those South and Central American countries that speak Spanish today, along with what is now the western United States and up the Pacific coast as far as Vancouver.

Florida's Northern Border

When England chartered the colony of Georgia in 1732, its border with Spanish Florida extended only to the Altamaha River, which empties into the ocean near the present-day town of Brunswick. But Spain claimed possession of all the land up to the Savannah River—the southern boundary of the Carolina Colony. (Figure 54)

The dispute eventually erupted into war. Georgia's colonists defeated the Spanish in the Battle of Bloody Marsh on St. Simons Island in 1739, and in so doing ended Spanish claims north of the St. Marys River. To this day, the St. Marys River serves as the eastern end of the northern border of Florida.

While the St. Marys River provides a natural boundary between Georgia and Florida from the Atlantic Ocean to the Okefenokee Swamp, it is

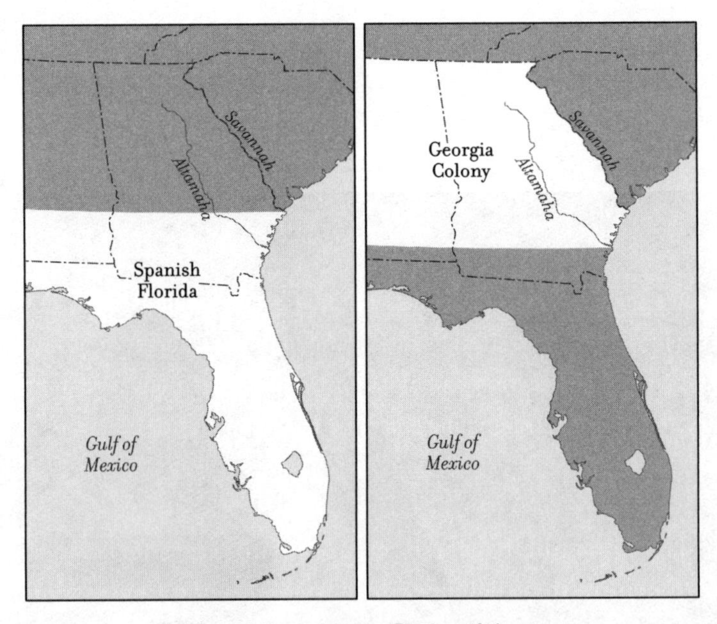

FIG. 54 Florida and Georgia's Conflicting Claims—1732

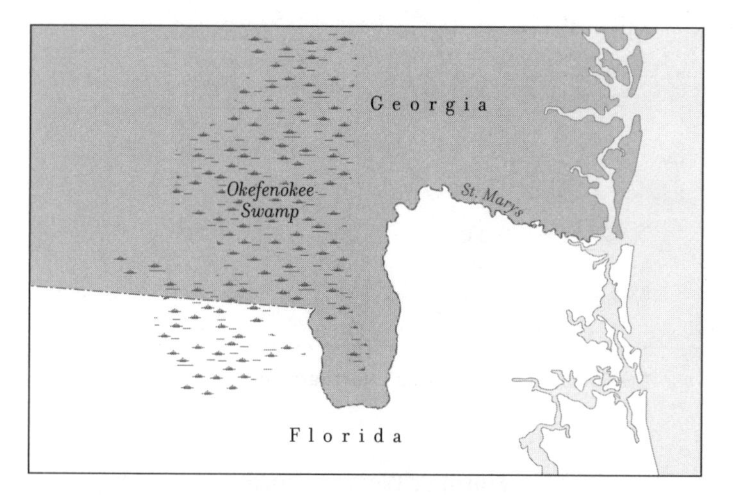

FIG. 55 **Battle of Bloody Marsh Border Adjustments—1739**

also a bit erratic. About 30 miles inland, the river takes a 90-degree turn to the south, then after about another 30 miles executes a U-turn. This accounts for the irregular jog in Florida's northeast corner. (Figure 55)

Since the residents of the Okefenokee Swamp were mostly alligators, birds, and bugs, the Georgians and Spaniards agreed that a straight-line border through the swamp would suffice. That line proceeds from the headwaters of the St. Marys River in the eastern side of the swamp westward to the convergence of the Flint River and the Chattahoochee River. Like the St. Marys River, this border, too, has remained in effect, a vestige of the uneasy relationship between 18th-century Spaniards and their colonial American counterparts.

But upon reaching this juncture, the border suddenly jumps 20 miles up the Chattahoochee River, then heads due west. (Figure 56) What happened here?

This leap to the north is an artifact that is older than the boundary agreed upon between Spain and the British colonists at Georgia. The line that resumes 20 miles north follows the 31st parallel, as specified in the first royal charter creating the Carolina Colony in 1663, nearly seventy years before the founding of Georgia. These boundaries are today the northern border of Florida.

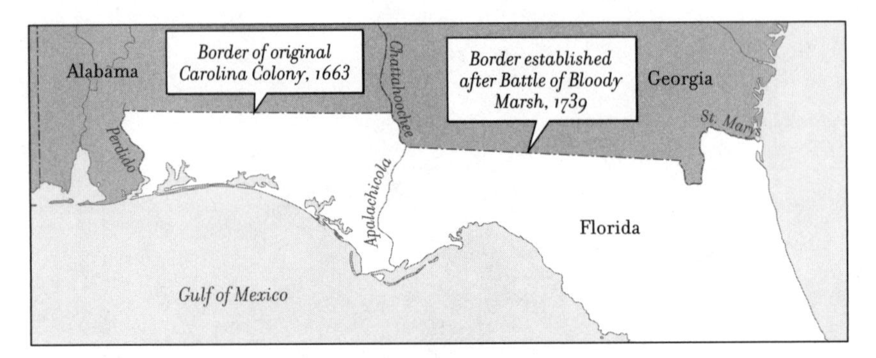

FIG. 56 The Components of Florida's Northern Border

Florida's Western Border

Originally, the western border of Spanish Florida was the Mississippi River. Not long after the American Revolution, that changed. In 1810, marking a further sign of Spain's weakening power, the young United States seized the westernmost portion of Spanish Florida, a chunk of land that extended from the Mississippi River to the Pearl River. Having purchased the western side of the Mississippi in the 1803 Louisiana Purchase, the new nation felt it was vital that it possess both sides of the river to ensure unchallenged access to the Gulf of Mexico and the sea.

A second seizure of land in western Florida took place in 1813, justified on the basis of Spain's support of the British in the ongoing War of 1812. This time the Americans took the adjacent chunk of land, eastward

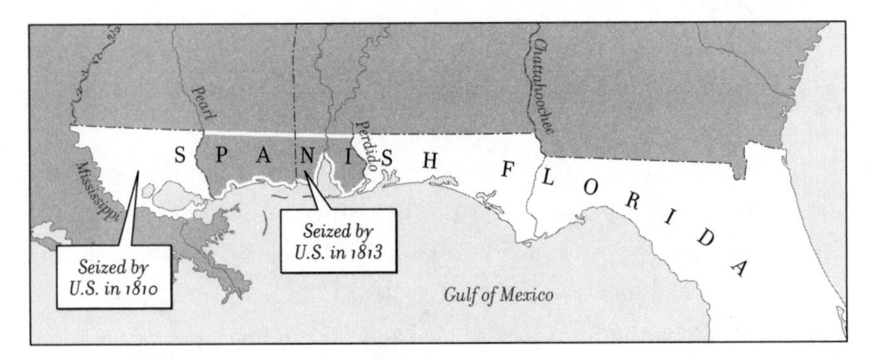

FIG. 57 American Acquisitions from Spanish Florida

to the Perdido River. This seizure included Mobile Bay, providing the United States a valuable port. (Figure 57) Today's western edge of Florida remains the Perdido River, just west of Pensacola.

Over the next ten years, Spain would lose virtually all of its possessions in the Americas. Recognizing the need to retreat, Spain released its remaining claims to Florida to the United States in the Adams-Onis Treaty (1819).

Why does Georgia share the same straight-line northern border with Alabama and Mississippi? And how come the straight segment of Georgia's western border doesn't simply point north and south?

Georgia's Eastern Border

Georgia was created from land originally included in the royal charter creating the Carolina Colony. (See Figure 151, in SOUTH CAROLINA.) But in 1732, King George II separated the land west of the Savannah

River to create Georgia. To this day, the Savannah River remains the border between Georgia and South Carolina, to the east.

Georgia's Northern Border

The charter creating Georgia stated that from the headwaters of the Savannah River, the colony's northern boundary was to be a direct line west to the Pacific Ocean. The fact that, at the time, France held claim to the land between the Mississippi River and the Rockies and Spain held claim to the land from the Rockies to the Pacific Ocean may have had something to do with the fact that Georgia limited its western claim to the Mississippi River.

A more realistic border problem stemmed from the fact that one can debate where exactly the source of the Savannah River is, given that rivers are an ingathering of various branches. By virtue of an agreement with the Cherokee Indians, the Chattooga River was chosen over the Hiwassee as the northernmost tributary of the Savannah.

But the headwaters of the Chattooga River are farther north than the 35th parallel, which was the boundary that North and South Carolina had established when the Carolina Colony divided in 1710. (For details, go to NORTH CAROLINA.) Hence, a line due west from the juncture of the Chattooga River and the 35th parallel became the northern border of Georgia.

Having the 35th parallel for a border seems simple enough, and indeed Georgia's northern border resides there to this day. But for more than a decade, the ground around that boundary was soaked with blood and tears. The underlying cause of the troubles was the realization, in the late 1700s, that the original surveyed line locating 35° was 12 miles north of where it should have been—according to North Carolina. Georgia disagreed with how far off the line was and, in any event, claimed the land was still theirs because it had been considered part of Georgia at the time of statehood. However, since this strip of land was part of a larger region belonging to the Cherokee, the issue wasn't urgent.

Until 1798. In that year, the Cherokee acceded to the first of a series of treaties through which they were ultimately forced to relocate in what is now Oklahoma. This evacuation came to be known as the Trail of Tears.

Following the 1798 treaty, in which the eastern end of the 12-mile-wide strip no longer belonged to the Cherokee, Georgia reasserted its claim to the land. North Carolina maintained that the boundary was the *actual* 35°, not the mistaken 35°, and that therefore it was entitled to the 12-mile-wide strip. Even South Carolina joined in, arguing that it was entitled to the land because it would originally have been part of South Carolina had the Cherokee not possessed it at the time the Carolina Colony split in two. (Figure 58)

For the next eight years, the issue remained unresolved. (According to Georgia, it remains unresolved even now.) Surveys were performed, disputed, and redone. During this period of uncertainty, the mountainous strip of land rapidly attracted residents looking for a neighborhood with a lot of hiding places and not a lot of law. In 1803, Georgia took matters in hand and organized the land as Walton County. Law and order were restored. But if Georgia had hoped this action would bolster its claim, it was wrong. It only made the land more appealing to North Carolina.

In 1808, Congress declared that the northern border of Georgia is the

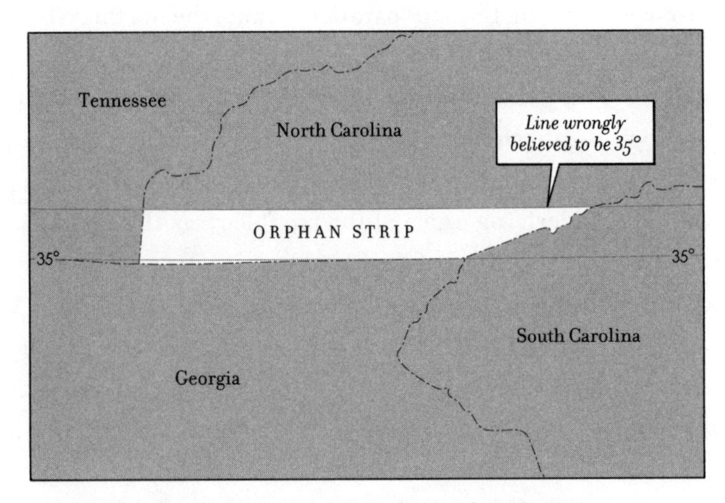

FIG. 58 The Orphan Strip—Disputed by Three States

actual 35° and that the disputed land, now known as the Orphan Strip, belonged to North Carolina. From Georgia's point of view, Congress was violating the constitutional provision by which the boundaries of the states cannot be altered without their consent. North Carolina sought to obtain that "consent" by sending its militia into the Orphan Strip in 1811. Georgia's militia fought back in two bloody battles and numerous skirmishes. But in every instance, the Georgians were outmatched. Within a few months, Georgian forces had been ousted from the Orphan Strip, and there the matter rests to this day.

Georgia's Southern Border

The origin of Georgia's southern border is simpler than that of its northern border. And bloodier. The colony's royal charter defined its southern border as the Altamaha River, some 45 miles *north* of the boundary described in the 1663 charter of its parent colony, Carolina. (Figure 59) One suspects that King George II, while expanding England's settlements southward, was also expressing caution, if not deference, toward Spain, which was wary of British encroachment on Florida.

The Georgia Colony was an ideal buffer between the thriving settlement at Charleston and the Spanish settlement at St. Augustine. Indeed, conflict did arise. Spain very much distrusted (rightfully, as it turned out) the continued growth of England's colonial holdings. In 1742, war

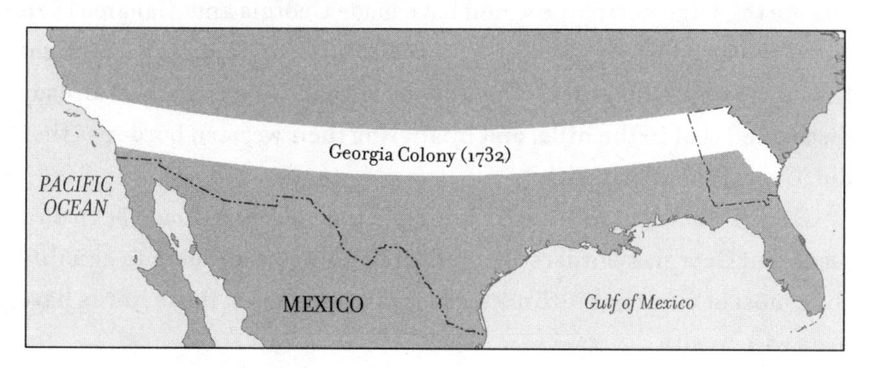

FIG. 59 Georgia According to 1732 Charter

broke out between Spain's Florida and England's Georgia, culminating in Georgia's victory at the Battle of Bloody Marsh on St. Simons Island. In the agreement that followed, the border between Georgia and Florida was established as the St. Marys River from its mouth at the ocean to its source in the Okefenokee Swamp. From that point, the border continued west as a straight line to the mouth of the Chattahoochee River. This border remains to this day. (See Figure 55, in FLORIDA.)

Georgia's Western Border

Shortly after the American Revolution, Georgia joined with the other states that had land claims west of the Appalachians and donated that land to the federal government. In the case of Georgia, the donated land would later become Alabama and Mississippi. Since there were no Appalachians except at the northern end of Georgia, the Chattahoochee River served as the divide. The Chattahoochee made an ideal border because it was located in such a way that Georgia's colonial claims could eventually be divided into three nearly equal states (Georgia, Alabama, and Mississippi).

One drawback to the Chattahoochee as a boundary was that midway up to Georgia's northern border, the river turns northeastward. For that reason, at the point where the Chattahoochee turns, the river border was replaced with a straight line.

But it was a straight line that angled slightly to the west rather than due north. A due north line would have made Georgia and Alabama even closer in size than they are now. Why, then, did Georgia angle the straight line of its western border? The answer is coal. Georgians had already discovered coal in the hills, and by angling their western border as they did, Georgia kept that coal.

Georgia's borders reflect the principle that *all states should be created equal.* But Georgia's borders also reflect that its commitment to equality (like most of ours) has its limits. More often than not, those limits have to do with wealth.

HAWAII

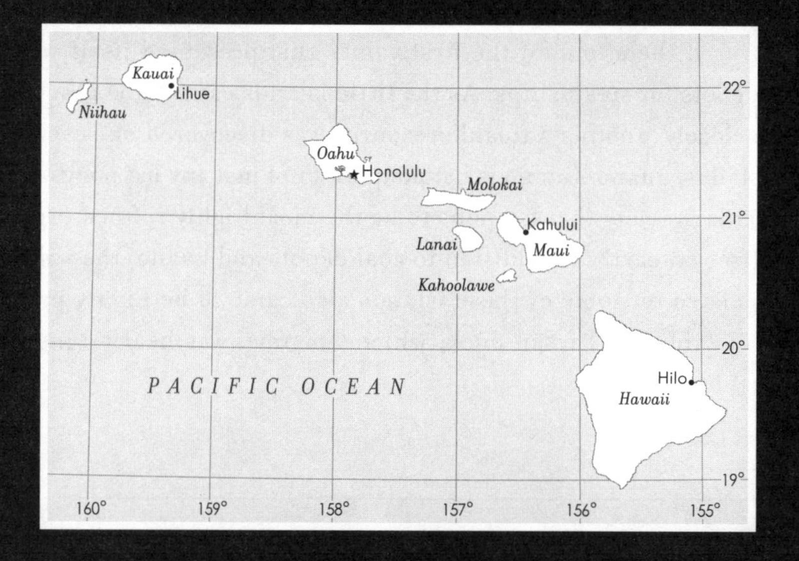

Aren't the reasons for Hawaii's borders pretty obvious? (Actually, not entirely.)

The reasons for Hawaii's borders are not quite as obvious as one might think. The boundary of the Hawaiian Islands is, obviously, the Pacific Ocean. And those islands consist of Hawaii, Maui, Kahoolawe, Lanai, Molokai, Oahu, Kauai, and Niihau. And some maps show Kaula.

And other maps show some other islands. The fact is the state of Hawaii extends over 1,000 miles northwest beyond these main Hawaiian islands. (Figure 60) Virtually all of these additional thousand miles consists of small, unpopulated islands, atolls, submerged banks, and reefs. Why are these landforms included within the borders of Hawaii?

When these smaller islands were first discovered by the Western world in the mid-19th century, they were quickly recognized for the ways in which they could be of value. Since it was ship captains who discovered them, one of the first values that presented itself was as coal depots for steamships. As the little islands came to be inspected more closely, another valuable resource was discovered on several of the islands: guano. Guano is bat poop. But not just any bat poop. It is a particular variety that happens to be the most highly refined organic fertilizer on earth. In addition to coal depots and guano, the wildlife that existed on some of these islands also came to be highly prized, most notably the Laysan duck, which, in time, was nearly rendered extinct by hunters.

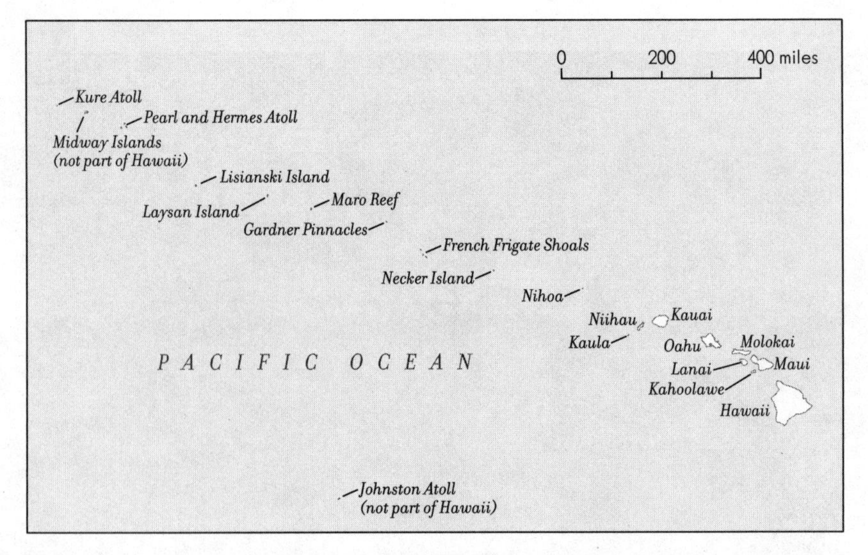

FIG. 60 The Full Extent of the State of Hawaii

Hawaii's American Border

The United States had a particular interest in the Hawaiian Islands at the time of their discovery, since, unlike England and France, the United States possessed no overseas colonies. In the Far East, American shipping was dependent on other nations for refueling. The United States was also without a naval base from which it could protect American commerce. It is hardly a surprise, then, that immediately after the Civil War, which had presented more urgent military demands, the United States laid claim to a small island in the uninhabited western end of the Hawaiian chain. The United States used the island as a coal depot. Since it was situated halfway between the west coast of the United States and China, it came to be known as Midway.

The royal rulers of Hawaii, meanwhile, set about establishing official boundaries to define their island chain. In 1854, they officially declared to Great Britain, France, and the United States that the islands of Hawaii, Maui, Oahu, Kauai, Molokai, Lanai, Niihau, Kahoolawe, Nihoa, Molokini, Lehua, and Kaula, were all within the domain of Hawaii. By 1886, Laysan, Lisianski, Palmyra, and Kure had been added to the Hawaiian domain.

The commercial value of these islands proved to be both a blessing and a curse to the Hawaiian government. The Hawaiian government profited from leases it issued to foreign corporations (most often American) for tracts of land. But introducing a foreign culture into the social ecology of so small a nation resulted in fundamental change. In 1893, Hawaii's queen, Liliuokalani, surrendered her kingdom, under protest, and elections were held in what was now the Republic of Hawaii. This event was hardly an internal democratic revolution. Hawaii's first president was Sanford Dole, brother of the founder of the Hawaiian-based Dole Pineapple Company. Seven years after the local American industrialists had taken control of the country, the United States declared Hawaii a U.S. territory. In 1959, that territory became the nation's fiftieth state.

Hawaii's Borders Within the Larger American Boundary

Not all the islands in this Pacific Ocean cluster are within the borders of the state of Hawaii. Midway, for example, despite being the island first declared an American possession, is not part of the state of Hawaii, even though it is not as far-flung as Kure, which is within the state's border. Johnston Island, too, to the south of the cluster, is not considered to be within the boundary of the state.

Midway, valued since the mid-19th century as a refueling hub for steamships and, later, airplanes, was transferred to the jurisdiction of the United States Navy in 1903. During World War II, it was the scene of a critical battle in America's struggle to defeat Japan. Because Midway had remained entirely devoted to military defense, it continued to be a preserve of the U.S. government when Hawaii became a state. It remains, to this day, a military base.

Johnston Island's exclusion from the Hawaiian boundary has been more controversial. In the mid-19th century, both the United States and Hawaii periodically hoisted their respective flags on the island, which was valued for the guano it contained. In 1926, President Calvin Coolidge declared Johnston Island to be a bird sanctuary, to be maintained by the Department of Agriculture. (Unfortunately, the Department of Agriculture had no ships.) Why then, when Hawaii became a state in 1959, was Johnston Island not included in its borders? The answer has to do with a series of atomic bomb tests the United States conducted in the area of Johnston Island. After that, the island was truly for the birds.

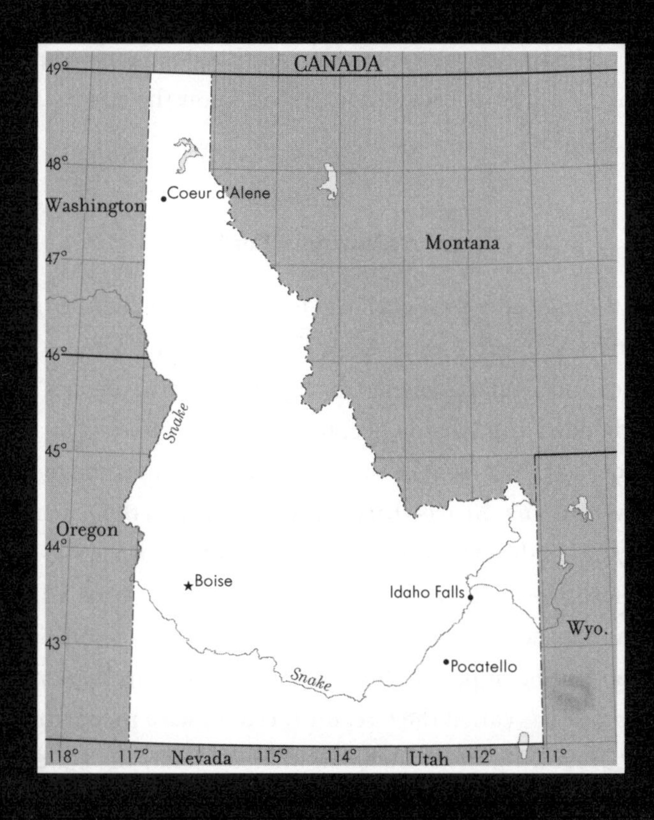

Why does Idaho have that little panhandle at the top? And why do its squiggly eastern and western borders suddenly go straight? Wouldn't it have been tidier, in a region of straight-line borders, for Idaho's eastern border to have simply been a continuation of its border with Wyoming? Isn't Montana big enough as it is?

Idaho's Southern Border

Idaho inherited its southern border from England and Spain. The two nations had contested the region that covers what is today northern California up to Alaska. The dispute was settled by dividing their interests along the 42nd parallel, which remains to this day as the southern border of Idaho. (To find out why they settled upon the 42nd parallel, see DON'T SKIP THIS.)

Idaho's Northern Border

Idaho's northern border first surfaced forty-five years before the creation of Idaho, and 1,000 miles away. After the Louisiana Purchase (1803), the United States and England needed to negotiate where America's newly acquired land ended and Canada began. In 1818, they agreed to a boundary along the 49th parallel, from Lake of the Woods, Minnesota, to the crest of the Rocky Mountains. Nearly thirty years later, when the American/British partnership in the Oregon Country came to an end, the two nations yet again extended the 49th parallel as a boundary, this time all the way to Puget Sound on the west coast. (For more details, go to OREGON.) The future Idaho had been part of the Oregon Country, and was now part of what the Americans called the Oregon Territory with its northern border at 49°. The Oregon Territory included what are today the states of Washington, Oregon, and Idaho, along with those parts of Montana and Wyoming west of the Continental Divide.

Idaho's Western Border

Idaho's western border emerged in stages. Its southern half first appeared in 1859, when Oregon, in becoming a state, released its eastern territorial region. Oregon's new eastern border was the Snake River south to the point where the Snake was joined by the Owyhee River, at which point a straight line due south formed the border. The discarded land to the east became part of the Washington Territory. (See Figure 170, in WASHING-

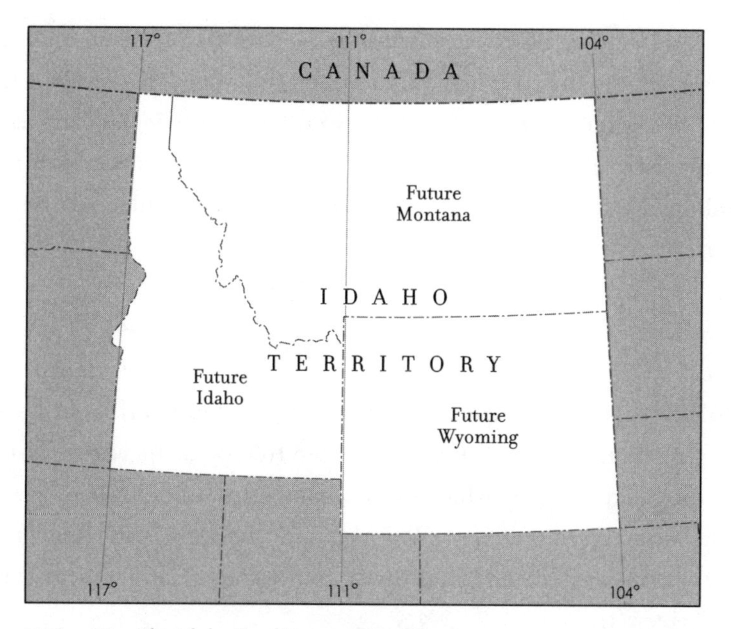

FIG. 61 The Idaho Territory—1863–1864

TON.) Four years later, the Washington Territory similarly shed its eastern half, and its discarded land then became part of the Idaho Territory. (Figure 61)

Mirroring Oregon, Washington also divided itself along the Snake River, in its case northward to the juncture of the Snake and the Clearwater rivers. From this point, a line due north took over. That line turns out to be the same longitude as that used by Oregon, 117°. Idaho's western border, therefore, comprised two straight lines along the same longitude connected by a segment of the Snake River.

But why had the territory of Washington rid itself of land where, in 1860, gold had been discovered?

The reason was that following the gold discovery, tens of thousands of miners flocked to the region around the Clearwater and Salmon rivers, above the city of Boise. Governing this sudden population of newcomers, a very different breed of individuals from those already settled in the Puget Sound region of Washington, was very difficult. The two groups were separated not only by the Cascade Mountains but also by cultural differences that added up to political differences. Indeed, so many newcomers were pouring into the gold regions that it became obvious that it was only a matter of time before they would outnumber the original settlers and begin electing themselves to the positions of power. Before that could happen, the Puget Sound–based population decided to divide the territory.

Idaho's Eastern Border

In 1863, when the Washington Territory separated itself from its gold-mining eastern regions, Congress created the Idaho Territory. This territory included not only what is today Idaho but all of what was later to become Montana and nearly all of what later became Wyoming. (Figure 61) Territories were often far larger than the states Congress envisioned creating from them.

In creating the borders of the Idaho Territory, Congress established a multistate boundary that survives to this day—despite the fact that it no longer has anything to do with Idaho! Starting at the Canadian border, a straight line along the 104th meridian stands out on the map as it divides the eastern edges of Montana and Wyoming from the western edges of North Dakota, South Dakota, and Nebraska. (To learn why this line is at 104°, as opposed to some other location, see DON'T SKIP THIS.)

It became clear very quickly, however, that these borders, even as a temporary measure, were not workable. The mountains in Idaho presented so formidable a barrier that those who lived on one side had little, if any, interaction with those who lived on the other. One year after Congress created the Idaho Territory, it divided it into Idaho and Montana. Since the enormous mountain barrier was the key difficulty, Congress created a border along the mountain crests.

But why did it choose the mountains it did? The Idaho legislature proposed that the border follow the Continental Divide—water east of the Divide makes its way to the Atlantic Ocean; water west of the Divide makes its way to the Pacific Ocean. (Figure 62)

But this is not the Idaho we know. The middle of Idaho's eastern bor-

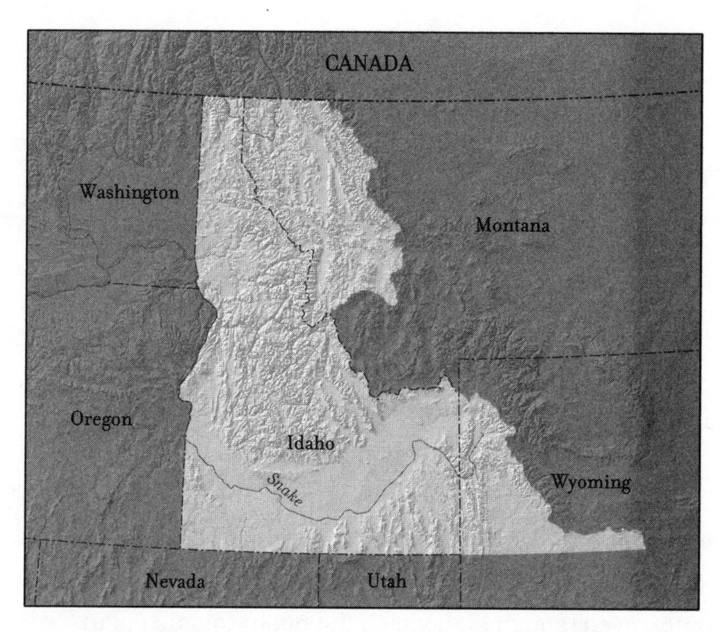

FIG. 62 The Proposed Eastern Border for Idaho Along the Continental Divide

der does follow the Continental Divide. But the upper and lower segments of its eastern border do not. Why did Congress alter such a seemingly sensible solution?

The lower segment of Idaho's eastern border is a straight line, due south, at 111° W longitude. (Figure 63) Why? Congress had two reasons for employing this line of longitude. First, the Snake River valley, a wide fertile swath, extends just about exactly to 111° W longitude. By placing the border there, Congress was preserving for otherwise mountainous Idaho the agricultural valley of the Snake River. Second, by locating Idaho's eastern border at 111°, Congress was also locating the western border of Idaho's future neighbor to the east. Since Congress had previously established the eastern border of the short-lived Idaho Territory at 104° W longitude, this future neighbor would have seven degrees of width. And it does—that state turning out to be Wyoming. Because of these two borders—111° and 104°—established by Congress before Wyoming even existed, Wyoming came to have the same width as North Dakota, South Dakota,

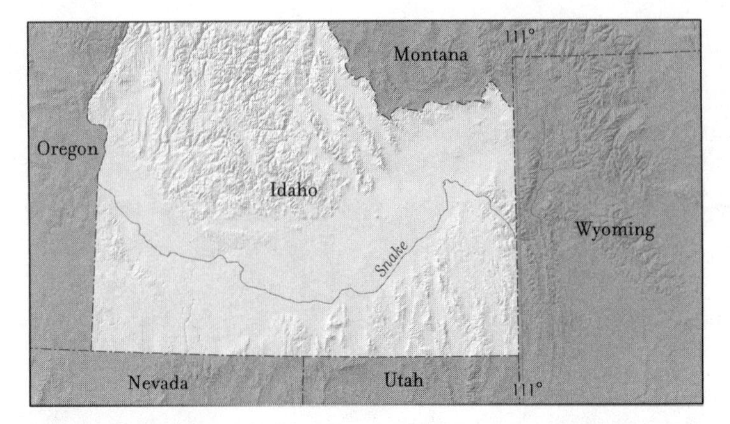

FIG. 63 Idaho's Eastern Border—Straight-Line Location

Colorado, Washington, and Oregon. Idaho's eastern border, then, was located where it was to accommodate a larger plan for the creation of equal-sized states, even though Idaho itself did not fit into that plan.

Less geometric motives influenced the northern segment of Idaho's eastern border, the section that became the skinny panhandle. When the Idaho Territory was created, former Ohio congressman Sidney Edgerton relocated to the territory, where he had obtained a judicial appointment. Upon arriving, Edgerton discovered that the governor had assigned him to an outlying district east of the Rockies. Edgerton felt snubbed.

The following year, when Idaho proposed dividing its territory, the residents east of the mountains chose Judge Edgerton to represent them in the creation of what became Montana. Not only was he a former congressman, he was personally acquainted with President Abraham Lincoln. Edgerton went to Washington with $2,000 in gold packed away. Somehow, he derailed Idaho's proposed Continental Divide boundary, pushing the line back to the crest of the Bitterroot Mountains. Thus, Idaho's eastern border is, in part, an enduring monument to the fact that a single person can change the course of history. But the person has to know how to pack.

Edgerton's efforts, however, do not explain the straight-line segment up at the top of Idaho's eastern border, since that line is clearly not

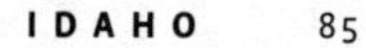

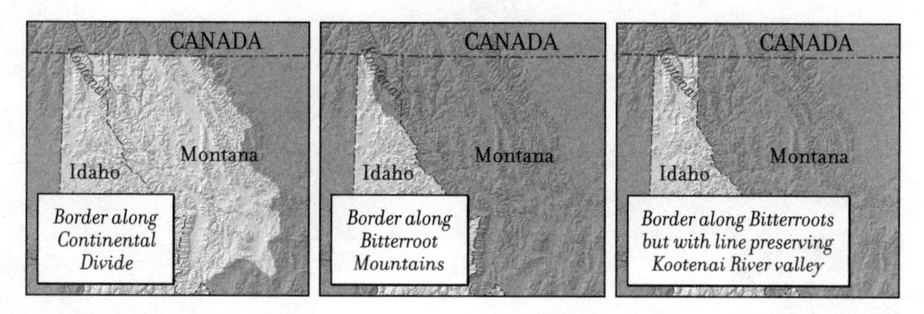

FIG. 64 Idaho's Panhandle and Alternative Proposals

following a mountain crest. Why didn't Congress stick with the mountain range? (Figure 64)

By locating this straight-line border where it is, it preserves for Idaho valuable agricultural land in the Kootenai River watershed. Idaho's eastern border is, ultimately, an enduring monument to the fact that an individual can change the course of events only up to a point.

ILLINOIS

Why is Illinois' straight-line northern border located so much farther north than that of its eastern neighbor, Indiana? And how come Illinois' squiggly eastern border suddenly goes straight? Why not just stick with the river?

More than that of any other state, the shape of Illinois preserves the boundaries of the land won by British and American forces in the French and Indian War, 1754–1763. (See Figure 4, in DON'T SKIP THIS.) Illinois itself first appeared in 1809, when the United States began to carve states from that land (which had come to be known as the Northwest Territory), pursuant to its Northwest Ordinance of 1787.

Illinois' Western and Eastern Borders

The borders at the bottom of Illinois preserve the southernmost boundaries of the Northwest Territory. The Mississippi River is the western border of Illinois, just as it had been the western border of the land won from France. The Ohio River, also a boundary from the victory in the French and Indian War, even now forms the southeastern border of Illinois. And the point at which the Ohio River empties into the Mississippi River is the southernmost point of Illinois, just as it had been the southernmost point of the land won in the French and Indian War.

In creating Illinois and Indiana, the Wabash River was an ideal boundary, since it would divide the two territories almost equally along a north-south line. Up to a point. That point is where the Wabash wanders off to the east. For this reason, the Northwest Ordinance called for a straight line to take over as the border from Vincennes to Lake Michigan. But Illinois' eastern border continues up the Wabash for more than 40 miles beyond Vincennes before the straight line takes over. Why? (Figure 65)

Had the straight line been located at Vincennes, the meandering Wabash would have crossed it several times, creating "islands" of jurisdiction across the river from their respective states. Neither Indiana nor Illinois needed that headache. Therefore, when Congress created the actual boundaries of Illinois, it commenced the straight line at the northernmost point that the Wabash crosses the longitudinal line on which Vincennes is located.

Illinois' Northern Border

Illinois' northern border also departed from that specified in the Northwest Ordinance, only this time not as an act of mutually consenting

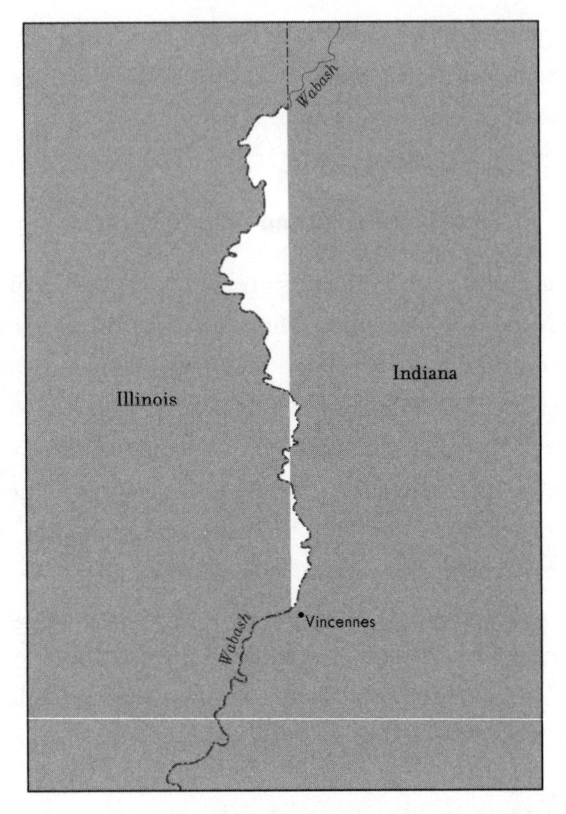

FIG. 65 Potential "Islands" of Jurisdiction in the Illinois/Indiana Border

states. Among the territorial borders described in the Northwest Ordinance was "an east and west line drawn through the southerly bend or extreme of Lake Michigan." (Figure 66) One problem with this border was that any state below the line would have no window on Lake Michigan and the important transportation network provided by the Great Lakes. In fairness to Thomas Jefferson, the primary architect of the Northwest Ordinance, in 1787 the Great Lakes would not be considered a major transportation network for another twenty years or so, when the idea of a canal connecting the lakes to the Hudson River (and thus to the Atlantic Ocean) began to take shape.

Consequently, when the residents of Illinois decided to seek state-

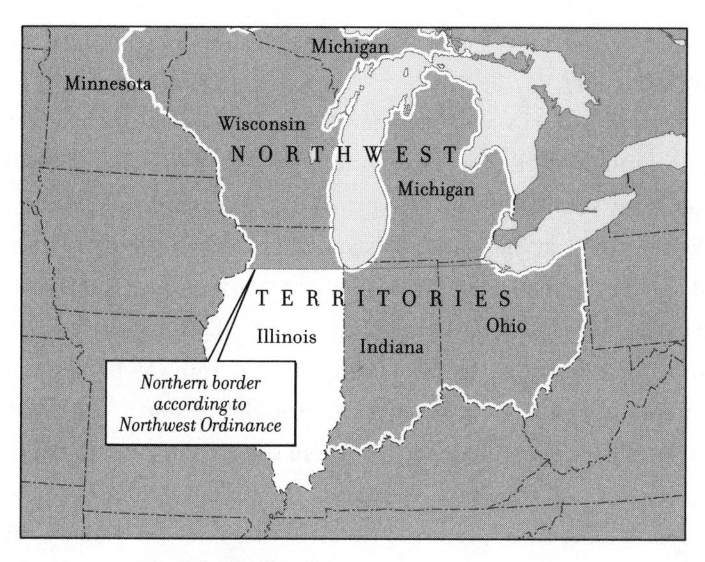

FIG. 66 The Original Illinois

hood in 1817, they now knew just how critical access to Lake Michigan would be to the economy of the state. In that same year, construction began on the Erie Canal. For this reason, Illinois sought to have its northern border adjusted to provide the state with a window on Lake Michigan. Congress proposed locating the northern border of Illinois at the same latitude as Indiana's—Indiana having similarly adjusted its border a short time earlier by having it relocated ten miles north of the southernmost point of Lake Michigan. But the Illinois statehood delegation urged Congress to locate its border nearly 60 miles north! And they succeeded. Why?

When Illinois made its bid for statehood, Missouri was also becoming a state. Missouri sought admission to the Union as a slave state, whereas by law none of the states created from the Northwest Territory could have slavery. The drift toward Civil War was already a conscious concern, as revealed by Illinois statehood delegate Nathaniel Pope's observation that a new state connected to New York would afford "additional security to the perpetuity of the Union." What in the world was this man's logic? Illinois was not even close to New York.

Plus, what did this have to do with Missouri? Not to mention Illinois' northern border?

As far as Missouri was concerned, the fear for the future security of the Union included the fact that so many of the nation's western rivers find their way to the Missouri River, which, in turn, finds its way to the Mississippi River at St. Louis, Missouri. This network of rivers represented a vast system of transportation for resources, and those resources all led to the slave-holding state of Missouri and points south.

Illinois also borders the Mississippi River, and was poised to play a similar role in channeling resources through the south. But it also had the option of directing the resources from the rivers in northern Illinois to Lake Michigan. From Lake Michigan, the goods could proceed to Lake Huron, then Lake Erie, then into the Erie Canal to the Hudson River, at the mouth of which is Manhattan and access to the sea. (Figure 67) This connection was what Nathaniel Pope was referring to when he linked Illinois to New York. And this connection was why Illinois could contribute to the security of the Union. Still, why did Illinois need to continue 60 miles to the north?

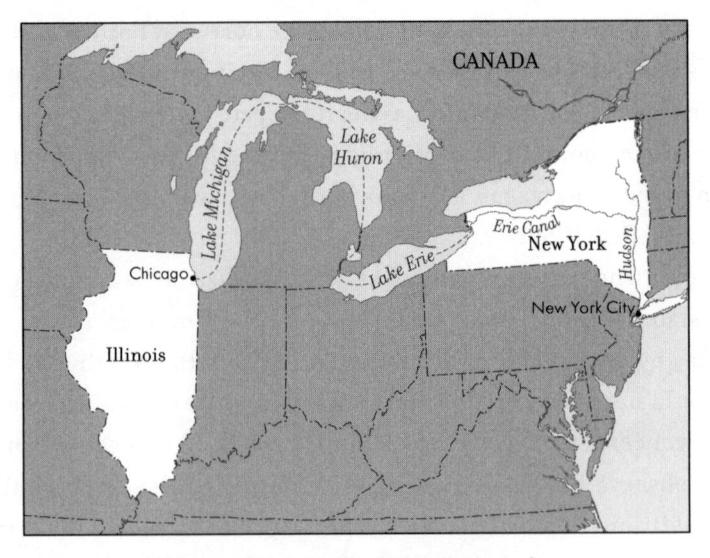

FIG. 67 Northern Illinois' Connection to New York

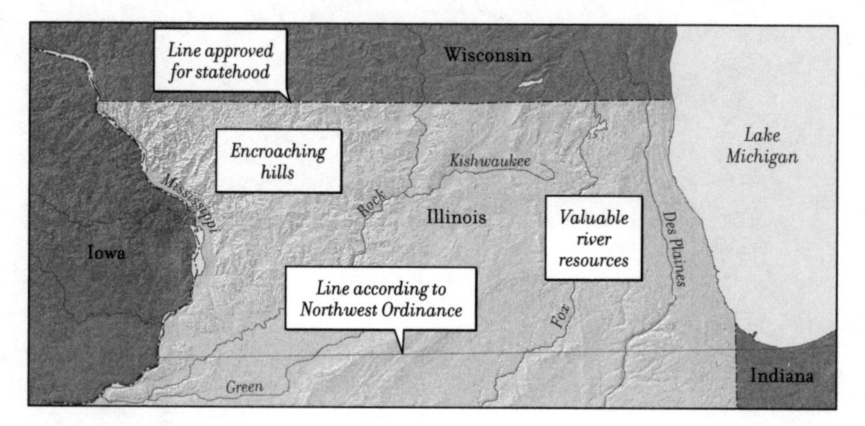

FIG. 68 Northern Border of Illinois

The canals Illinois needed in order to divert commerce from the Mississippi River to Lake Michigan would draw from a region of rivers and tributaries almost as far as 60 miles north of the line described in the Northwest Ordinance. Farther north than that, hills begin to dominate the landscape, and hills make canal construction far costlier. (Figure 68) If Illinois had not gained possession of the flat land leading up to the hills, the canal promoters could have lost valuable real estate, possibly jeopardizing the venture.

INDIANA

How come Indiana's northern border doesn't line up with its neighbors'? Why are Indiana's straight lines located where they are? And how come Indiana got to border more of the bottom of Lake Michigan than did Illinois?

Indiana was previously part of the land that came into British possession (and, after the Revolution, American possession) following the French and Indian War, 1754–1763. (See DON'T SKIP THIS.) In first emerged as a territory in 1800. It consisted then of the area west of a line that divided the eastern third of the region, where settlers were then populating, from the rest of the region. As the name Indiana implies, Indians occupied this western region. But not for long. (See Figure 135, in OHIO.)

Indiana's Eastern Border

The boundary used to separate the eastern third of the territory, known as the Ohio Territory, from the Indiana Territory was that specified in the Northwest Ordinance of 1787: a line from the juncture of the Ohio and Great Miami rivers due north to the Canadian border. The southern half of this first American territorial division remains to this day as the eastern border of Indiana.

Indiana's Southern Border

The land acquired in the French and Indian War comprised a kind of triangle, with the Mississippi River, the Ohio River, and the Great Lakes as its "sides." (See Figure 4, in DON'T SKIP THIS.) A remnant of these boundaries can be found in Indiana's southern border, which remains to this day the Ohio River.

Indiana's Western Border

The rush of settlers populating the Ohio Territory soon began populating the Indiana Territory as well. In 1805, Congress further divided the Indiana Territory, as it had anticipated in the Northwest Ordinance. That act stipulated that the Wabash River would form the lower half of Indiana's western border. The Wabash was an ideal border because it divided Indiana and the newly created territory of Illinois just about evenly—that is, until it veers to the east. Consequently, the law called for a straight line to take over at Vincennes, heading due north to the state's northern border. But the straight-line segment of Indiana's western border does not commence at Vincennes. It commences more than 40 miles north at what appears to be nowhere in particular.

In fact, it is somewhere very much in particular. It is the northernmost point at which the Wabash River crosses the longitude of Vincennes. By commencing the straight line here, Congress preserved the same width that would have been achieved had the line commenced at Vincennes, but avoided creating isolated pockets where the Wabash would have crossed back and forth over the straight line as the river meandered east and west, resulting in "islands" of jurisdiction isolated by the river. (See Figure 65, in ILLINOIS.)

Indiana's Northern Border

Initially, the northern border of Indiana was that specified in the Northwest Ordinance, a straight east-west line that intersected the southernmost point of Lake Michigan. But such a line left Indiana with no actual access to the Great Lakes, which had acquired new, enormous importance since the Ordinance because of their connection to the Erie Canal and, via the canal, access to the Atlantic Ocean. Congress therefore adjusted Indiana's northern border, locating it 10 miles farther north, where it remains to this day. (To find out why Ohio's border just misses Indiana's, go to OHIO.)

IOWA

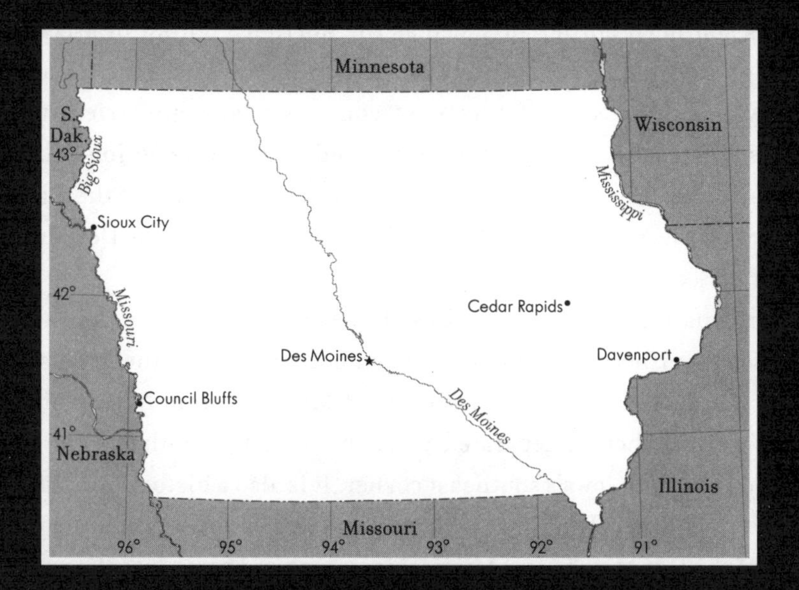

Why does Iowa have that little pointy piece on its southeast corner? Why are its straight-line northern and southern borders located where they are? And is Iowa's southern border *really* straight?

Iowa's Eastern Border

Viewed today, Iowa's shape hardly looks like something to die for, but in fact, Iowa went to war over its borders. The land that is now known as Iowa came into the possession of the United States as part of the

Louisiana Purchase (1803). With this acquisition, the eastern border of Iowa surfaced, that being the Mississippi River, which was the eastern extent of the Louisiana Purchase.

Iowa's Southern Border

When Iowa's neighbor to the south, Missouri, became a state in 1821, Iowa's southern border surfaced as the northern border of Missouri. Though it looks today like an innocuous line, the Missouri/Iowa border was actually one of the most controversial boundaries in the United States. As it was originally marked off by surveyor John C. Sullivan in 1816, the line went due north from the juncture of the Kansas River and the Missouri River for 100 miles, then east to the Des Moines River.

Terminating this line at the Des Moines River, at which point the Des Moines River becomes Iowa's southern border down to the Mississippi River, made sense because it prevented Missouri from having an "island" of jurisdiction separated from the state by a rather wide river. Hence the nib in Iowa's southeast corner. It is also a historic nib, known as the Half-Breed Tract. (Over the years, this region between the juncture of the Des Moines and Mississippi rivers had come into the possession of mixed-race offspring of Indians and whites.)

Sullivan's line had a problem, however. It wasn't straight. As it approached its eastern end, it curved northward. (Figure 69) Moreover, it

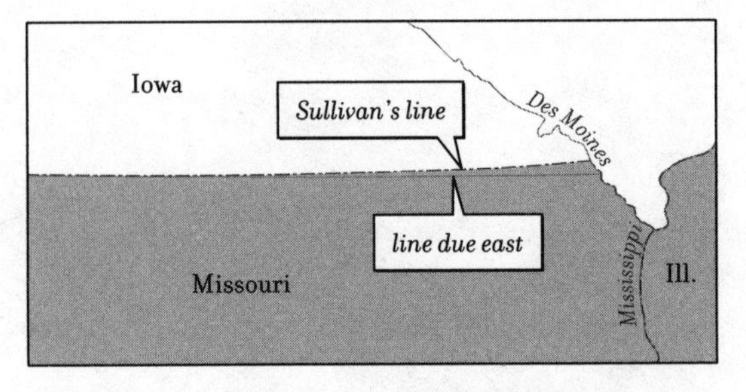

FIG. 69 Curvature of Sullivan's Line

did not clearly correspond to the boundary described in Missouri's first constitution, which defined its northern border as an east-west line on the same "parallel of latitude which passes through the rapids of the river Des Moines."

When Iowa applied for territorial status, it was well aware of the curvature in Sullivan's line. Missouri agreed to have the border resurveyed. The surveyor, Joseph C. Brown (hired by Missouri), drew an impeccable east-west line located at the latitude of the "Des Moines River rapids." This correction transferred 2,616 square miles of land . . . to Missouri! (Figure 70)

Iowans accused Missouri of inventing a new interpretation of the term, "Des Moines rapids," basing their accusation on the fact that *there are no rapids in the Des Moines River.* According to Iowa, the phrase "Des Moines rapids" was used by Frenchmen navigating the Mississippi River to describe the suddenly rapid current of the Mississippi as it narrowed approaching its juncture with the Des Moines River.

Meanwhile, a mystery emerged. Who had told Sullivan where to draw his line in the first place? In 1838, Iowa contacted the War Department,

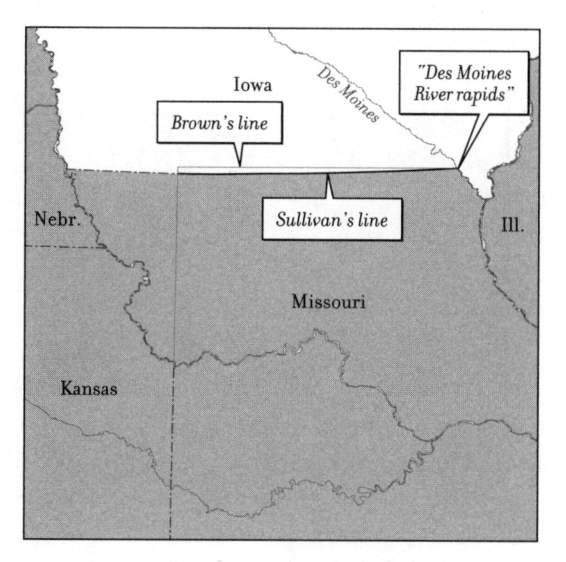

FIG. 70 Iowa/Missouri Border Dispute

which oversaw the Army engineers who surveyed the newly forming borders. The War Department reported back that it found

> no instructions were given by this department . . . respecting the running of the lines in question; but that Mr. William Rector, surveyor of the lands of the United States in Missouri and Illinois, had this done upon his own responsibility, and gave the necessary instructions.

What the War Department did not say was that Mr. William Rector—or, at the time he put Sullivan to work, *General* William Rector—had been dismissed from his post in 1824 because of charges of corruption and questions of competence.

Ultimately, the Supreme Court resolved the long-standing dispute. The court stated in 1849 that it was unable to ascertain what was meant by the phrase "the rapids of the river Des Moines." On the other hand, Sullivan's line clearly had several dubious elements in terms of its origin and execution. Still, because Sullivan's line had been the recognized border for so long, the Court ruled that it should remain the border between Iowa and Missouri.

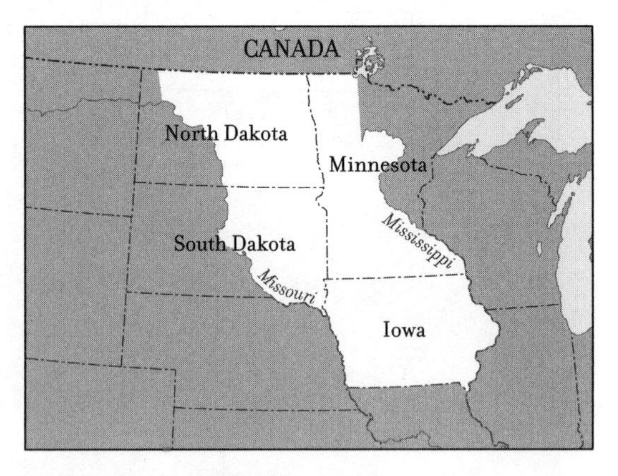

FIG. 71 **Iowa Territory—1838**

Iowa's Western Border

When Iowa was granted territorial status in 1838, it came with a vast amount of land—everything between the Mississippi and Missouri rivers, from present-day Iowa up to the Canadian border. (Figure 71) While these boundaries would eventually be redefined, nearly all of Iowa's western border had now emerged: the Missouri River. Upon becoming a state eight years later, the northernmost segment of Iowa's western border followed the Big Sioux River. Switching to this river resulted in a more rectangular shape, since the Missouri was veering increasingly westward.

Iowa's Northern Border

Territorial governor Robert Lucas proposed a northern border for Iowa extending to what is today Minneapolis. He believed Iowa's natural boundaries were the Missouri and Big Sioux rivers on the west, the St. Peter's River (now known as the Minnesota River) on the north, the Mississippi River on the east, and the Missouri border on the south. (Figure 72)

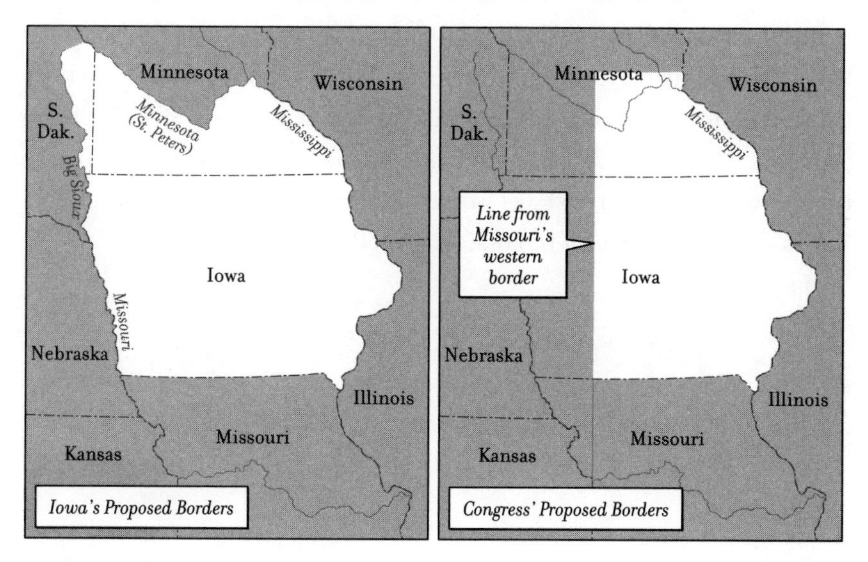

FIG. 72 Possible Iowas

Congress had a very different kind of Iowa in mind. Since the earliest days of the republic, a delicate balance had been maintained by the unwritten rule that for every slave state admitted to the Union, a free state would be admitted as well, and vice versa. Iowa, for example, was teamed up with Florida for admission to the Union.

But in admitting Texas (a slave state) into the union the year before, a provision was enacted allowing Texas, if it so chose, to divide itself into as many as five states. Northern states now sought to create as many states as possible from the territories in the prairies. In the case of Iowa, Congress envisioned a continuation of the midwest's vertical states, in the manner of Indiana, Illinois, Michigan, and Wisconsin. (Figure 72)

But Iowa voters rejected these boundaries. A new constitutional convention was organized, and in 1846, the delegates and then Congress approved a compromise boundary. This northern border cost Iowa its ambitious northern reach but restored the western border.

The northern boundary was now to be an east-west line located along 43°30' N latitude. But why 43°30'? There were two elements that led to this choice. One element was the fact that 43°30' was the optimum straight-line border for dividing those waterways that flowed northward to the Minnesota River from those that flowed southward to the Missouri River. (See Figure 97, in MINNESOTA.) The second factor wouldn't become evident until Congress created additional prairie states. By fixing the border at 43°30', Iowa has a height of almost exactly three degrees. In time, four other prairie states would also have three degrees of height: Kansas, Nebraska, South Dakota, and North Dakota.

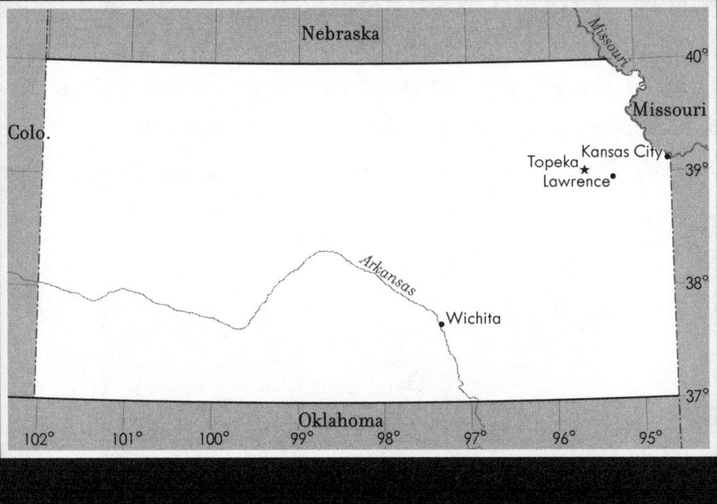

How come the southern border of Kansas doesn't line up with the southern border of its eastern neighbor, Missouri?—especially since Missouri's southern border comes from a long line stretching back east all the way to the Atlantic Ocean. And why does the Kansas line then continue west as the southern border of its neighbors, Colorado and Utah? On the other hand, how come the northern border of Kansas doesn't line up with those of either of its neighbors?

Not many states have borders more boring than Kansas. But looks can be deceiving. In the case of Kansas, its bland boundaries are the remnant of a fierce debate, one that is remembered today as "Bleeding Kansas."

All but the southwest corner of what is now Kansas came into American possession as part of the Louisiana Purchase (1803). Kansas got its southwest corner in 1846 when Texas, upon entering the Union and wishing to remain a slave state, ceded its land north of 36°30' as required by the Missouri Compromise. (See Figure 73. For more on the Missouri Compromise, see DON'T SKIP THIS.)

Ironically, the Missouri Compromise was later scrapped when Congress passed the Kansas-Nebraska Act. For this reason, the Kansas-Nebraska Act is most remembered for the role it then played in the nation's increasingly acrimonious debate over slavery. (The act also exerted a

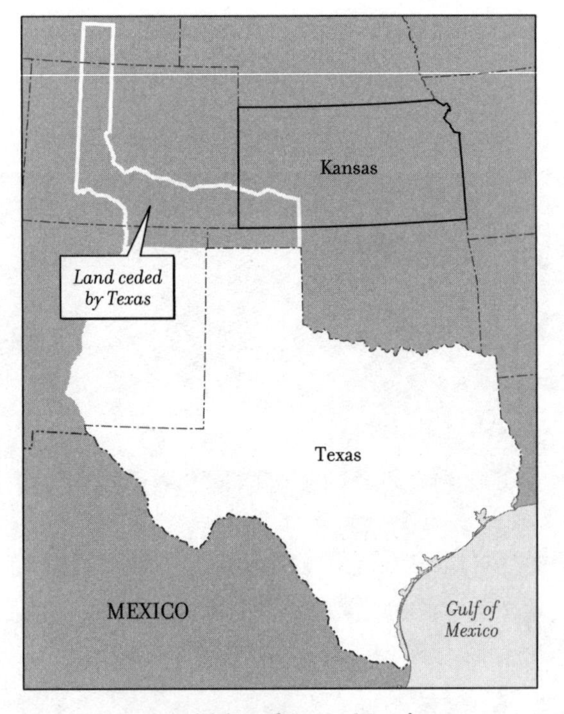

FIG. 73 Acquisition of Kansas' Southwest Corner—1846

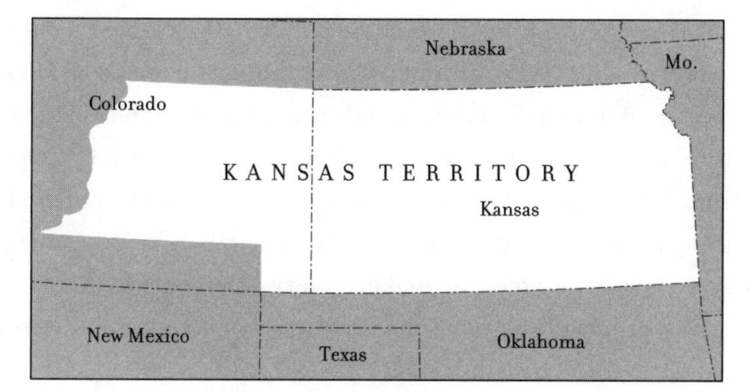

FIG. 74 The Kansas Territory—1854

major influence on future state borders, as we shall see in discussing Kansas's southern border.) In terms of Kansas, the primary importance of the legislation was that it created Kansas.

The initial territory of Kansas encompassed the present-day state, but extended west to what was then the New Mexico territorial border and, where that ended, west to the crest of the Rocky Mountains. (Figure 74)

Kansas' Eastern Border

When Congress created Missouri in 1821, the future state of Kansas acquired its first border. This eastern boundary of Kansas is the Missouri River down to its confluence with the Kansas River, at what is today the site of Kansas City. From this point, the border continues as a straight line due south to the southern end of the state. (To find out why it is a straight line located where it is, go to MISSOURI.) This border is the only border of Kansas that was not the subject of violent debate.

Kansas' Southern Border

When Congress began debate on the Kansas-Nebraska Act, the proposed southern border of Kansas was 36°30', thereby locating it on the northern border of Texas. Since Kansas would be north of 36°30', northerners assumed slavery would be prohibited there, under the terms of Missouri Compromise. But with the land recently acquired from the Mexican War (what would eventually become New Mexico, Arizona, Utah, Colorado, Wyoming, Montana, Idaho, and California), southerners realized that most of the country's future states were now north of 36°30'—enough to one day possess the votes to make slavery unconstitutional. Consequently, the proposal to create the territories of Kansas and Nebraska returned the conflict to center stage.

As a prerequisite to creating the two territories, southerners demanded a new agreement to replace the Missouri Compromise. Senator Stephen Douglas (who would go on to lose the presidential election of 1860 to Abraham Lincoln) proposed the solution—nationwide adoption of the concept of "popular sovereignty." It had been tried (with mixed results) four years earlier in the creation of the New Mexico Territory. Under popular sovereignty, the people of any given territory would decide for themselves whether or not to allow slavery.

The residents of Kansas then set about to persuade, verbally or violently, a majority of themselves to support one side or the other. For Kansas' sibling to the north, Nebraska, slavery was not a local issue, since the majority of its population opposed it. Those Nebraskans who did not oppose slavery tended to live in the southern end of the territory below the Platte River. Because of this, a movement commenced in Kansas among its pro-slavery citizens to make the Platte River the northern border of Kansas. (Figure 75) The added population this border would bring into Kansas would have given the pro-slavery faction the majority it needed under "popular sovereignty."

But Congress, not Kansas, determines state and territorial borders. And while a proposal was offered under which the northern border of Kansas would have been the Platte River and its south fork, anti-slavery forces in Congress rejected it, insisting on their own idea for Kansas' northern border.

Since popular sovereignty effectively repealed the Missouri Compromise, it also eliminated the significance of 36°30' as a border. As a result,

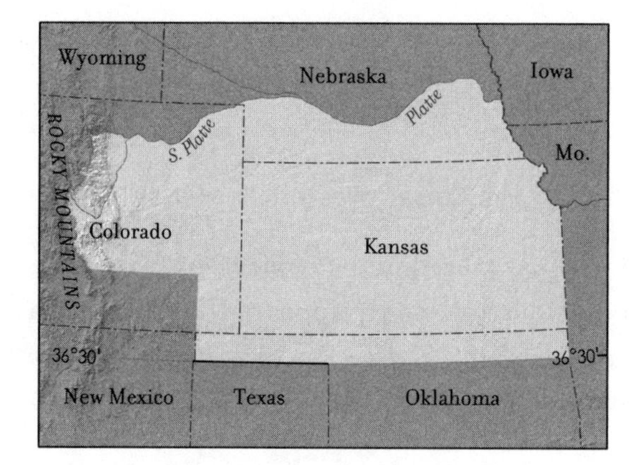

FIG. 75 **Kansas' Proposed Borders**

Congress made a slight but significant adjustment. It gave Kansas a southern border located at 37°—one half of a degree farther north. Why? The reason had to do with the northern border of Kansas that Congress insisted upon.

Kansas' Northern Border

Congress located the northern border of Kansas at 40°, exactly three degrees north of the southern border of Kansas that Congress had curiously adjusted by one-half of a degree. In doing so, Congress created the possibility that four states of equal height could be created from Kansas to the Canadian border. And, indeed, over the next thirty-five years, Nebraska, South Dakota, and North Dakota joined Kansas in forming a tier of prairie states, each having three degrees of height. (See Figure 11, in DON'T SKIP THIS.) Four years earlier, Congress had first used 37° as part of (and later *all* of) the northern border of the New Mexico Territory, enabling a similar tier of mountain states from New Mexico to Canada.

It may seem ironic that the borders of Kansas so clearly reflect the policy that all states should be created equal, since its borders also reflect Congress' washing its hands of the slavery debate and passing it to the states. Still, as the northern and southern borders of Kansas reflect, the effort for equality among the states (if not among humanity) prevailed. Kansas ultimately chose to prohibit slavery.

Kansas' Western Border

In 1858, gold was discovered in the mountains of western Kansas, and over 50,000 outsiders suddenly appeared in the hills and in towns that sprang up overnight. Native Kansans had to confront the possibility that retaining its gold mines might deprive them of their political control.

One year after the discovery of the gold, the newcomers in the mining region sought separate territorial status by proposing the "Territory of Jefferson." (See Figure 33, in COLORADO.) With the looming possibility of Civil War, and with the proposed Territory of Jefferson's tremendous wealth, residents of Kansas thought it likely that Congress would accede to the creation of this new territory. This fact, combined with concerns about maintaining political control, led Kansas to agree to separate from its gold mines in the west.

The miners proposed a border with Kansas located at 102° W longitude. But why there? Why not go a degree farther west and line up with the northwest corner of New Mexico? (Figure 76)

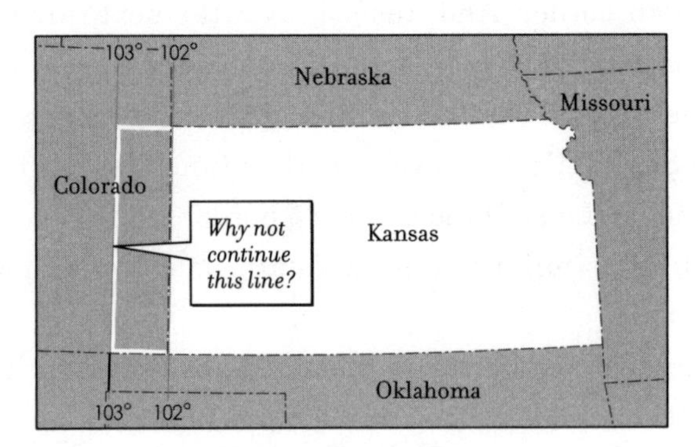

FIG. 76 Western Border of Kansas

One reason for choosing 102° was to give the new territory sufficient agricultural land east of the mountains. But another element was at work here, too. The proposed eastern and western borders for the proposed Territory of Jefferson (which Congress would rename Colorado) encompassed seven degrees of width. In time, five other western states—and much of Idaho—would also span seven degrees in width. (See Figure 13, in DON'T SKIP THIS.) Soon-to-emerge prototypes determined not only the height of Kansas but also its width in order that Colorado could fulfill its prototype, too.

Why does Kentucky's southern border suddenly sidestep to the south? How come Kentucky's southern border is the same (almost) as Virginia's? And since it's almost the same as Virginia's, why isn't it exactly the same as Virginia's?

Kentucky's Southern Border (or so it thought)

Prior to becoming a separate state, Kentucky was part of Virginia, who colonial boundaries were altered on numerous occasions. One of the alterations remains embedded in the boundaries of Kentucky to this day. In 1665, King Charles II fixed the border between the Virginia Colony and the Carolina Colony as an east-west line located at 36°30', the midpoint between the Chesapeake Bay and Albemarle Sound. Years later, that boundary went on to become the southern border of Kentucky.

Kentucky's Eastern Border

After the Revolution, some of the thirteen colonies, now states, were far larger than others, particularly given their colonial claims west of the Appalachian Mountains. These states agreed to release their western lands to the United States for the purpose of creating new states. A vision of equality among states was, however, only one factor in this decision. An additional factor, in the case of slave states such as Virginia, North Carolina, and Georgia, was the intention to create more slave states, to bring more pro-slavery votes to Congress.

Virginia's western lands eventually became two separate states, Kentucky and, later, West Virginia. Virginia's reason for holding on to what would become West Virginia is what accounts for the upper end of Kentucky's eastern border. The land that is now West Virginia borders Tug Fork, a branch of the Big Sandy River. And the Big Sandy River flows into the Ohio River, which flows into the Mississippi River, which flows to the sea. In the early 19th century, Virginia highly valued this access to the ocean from its western end. The land to the south and west of the Big Sandy River and the Ohio River was less critical to Virginia. This land became Kentucky. For this reason, Tug Fork, the Big Sandy River, and the Ohio River form the upper end of the eastern border of Kentucky.

The lower end of Kentucky's eastern border represents its line of division via the crest of the Appalachian Mountains from the present-day state of Virginia. But in some areas, the mountains that make up the Appalachians are oriented in such a way that a line connecting their crests would have resulted in a boundary so irregular that the land would have been difficult to govern. Therefore, to divide Kentucky from Virginia, Virginia employed other elements. At its southeast corner, the Cumberland Gap provided a natural gateway to Kentucky. With this as a starting point, the boundary followed the crest of the Appalachians northeastward until arriving at a point where the mountaintops cease to be oriented in a northeasterly way. Here a straight line continues northeastward until reaching Tug Fork. (See Figure 168, in VIRGINIA.)

Kentucky's Northern and Western Borders

Kentucky inherited its northern and western borders from, of all places, Francc.

Prior to the French and Indian War, all the land between the Ohio and Mississippi rivers was under the control of the French. Afterwards, the land came under British control. But the British prohibited the Americans from moving into the region, fearing that American expansion could weaken England's ability to control the colonies. After the Revolution, the area known as the Northwest Territory came under American control, but it was regarded as separate from the existing states. Thus, one of Virginia's (and soon, Kentucky's) borders became fixed at the Ohio River.

France continued to control the region west of the Mississippi River and east of the Rocky Mountains. In 1803, President Thomas Jefferson acquired the land in the Louisiana Purchase. Here, too, the land was regarded as separate from any of the existing states. Thus, the western border of Kentucky was fixed as its frontage on the Mississippi River.

Kentucky's Southern Border Revisited

No sooner had Virginia spawned Kentucky and, in similar fashion, North Carolina spawned Tennessee than the two cousins began squabbling over their mutual border. Virginia had commissioned Dr. Thomas Walker, one of its most renowned physicians, to survey the continuation of its southern border. Dr. Walker was, in fact, also a surveyor. He was also a merchant, a manufacturer, an explorer, a land speculator, and a public official. With Walker wearing so many hats, it is not surprising that the long east-west line he'd been hired to survey veered to the north before reaching its terminus at the Tennessee River. (Figure 77)

Over a hundred years of dispute ensued between Kentucky and Tennessee. In the conflict's opening round, the two states drew a compromise line in 1802. This line was located midway between Walker's line and 36°30', which should have been an extension of the southern boundary of Virginia. As it turned out, however, the Virginia border itself was erroneously located several miles to the north.

Yet another compromise was attempted in 1819, when General Andrew Jackson, acting on behalf of the U.S. government, purchased from the Chickasaw Indians lands bounded on the west by the Mississippi River, on the north and east by the Tennessee River, and on the south by the Mississippi state line. (See Figure 154, in TENNESSEE.) The Jackson Purchase, as it came to be called, was to be divided between Kentucky and Tennessee. But how? Tennessee agreed to accept a division of the Jackson Purchase at the actual 36°30', thereby placing the border for this newly acquired land several miles below what had long been accepted as its northern border.

The far western end of Kentucky is a small "island" that is separated from the rest of the state by the Mississippi River. This 11-square-mile

FIG. 77 Deviation in Walker's Line

area resulted from the fact that 36°30' crosses the meandering Mississippi several times, creating a pocket of land within its bend. Unlike Illinois, Utah, Mississippi, and Massachusetts, where state boundaries were adjusted to eliminate pockets of land with difficult access, this area was not similarly adjusted by making it part of Tennessee. Being a flood plain, the area was not conducive to harboring miscreants. And neither Kentucky nor Tennessee, having finally acquired a way to settle their longstanding border dispute, was anxious to spark a new one.

Nevertheless, a new dispute did erupt. In agreeing to the Jackson Purchase boundary, Kentucky accepted the Walker line as the rest of the boundary between the two states. Unfortunately, many of the original markers of the Walker line, mostly trees, were no longer standing. Consequently, the disputes continued until 1891, when the Supreme Court instructed the Army Corps of Engineers to survey a line that, as best as could be determined, followed the 1802 Compromise Line.

How come Louisiana's eastern border suddenly jumps across the Mississippi River? And why does it make this jump where it does? Why is Louisiana's straight-line northern border located where it is? And why does its western border suddenly go straight north rather than continue along the Sabine River?—wasn't Texas big enough already?

Louisiana was acquired by the United States in 1803 as part of the Louisiana Purchase. But the state we call Louisiana is only a small fraction o the land President Thomas Jefferson bought from Napoléon Bonaparte for $15 million. Exactly how small a fraction is unknown, since no one knew the borders of the land France sold us. What was known was that Spain's North American claims came right up to the Louisiana Purchase and several of these areas were under dispute. Two of these disputes directly affected the border of what would become Louisiana. One of them brought the United States and Spain to the brink of war.

Louisiana's Western and Eastern Borders

The Americans maintained that the southern end of the Louisiana Purchase included the land west of the Mississippi River to the Sabine River. This claim was based on the presence of French trading posts that had, at times, existed in this region. Spain, for its part, maintained that its claims extended as far as the Red River, along which it had established missions and which was an avenue of commerce that originated well within the realm of Spanish North America. (Figure 78) To underscore its claim, Spain began assembling troops across the Sabine River.

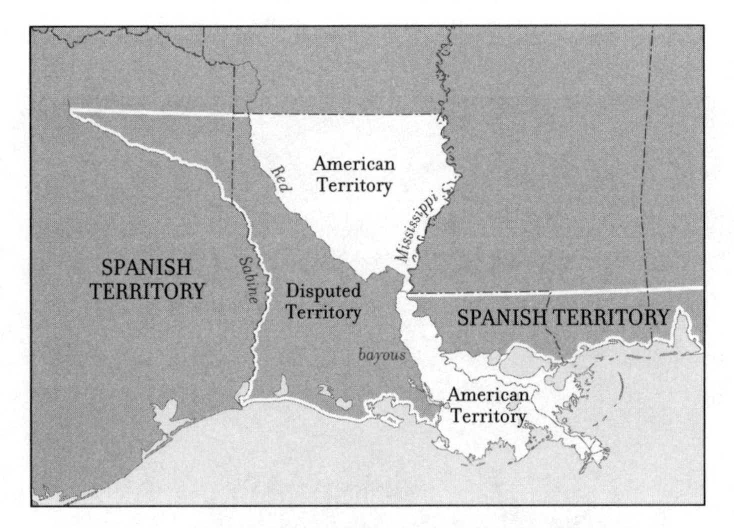

FIG. 78 Louisiana's Western Border—U.S./Spain Dispute

Fueling this fire was the fact that, to the east, Florida was a Spanish possession. And at the time of the Louisiana Purchase, the panhandle of Florida extended all the way to the Mississippi River and as far north as the 31st parallel. The United States, which viewed the Mississippi River as a vital economic artery, sought to possess both banks of the river.

In 1810, American forces seized the westernmost end of Florida—from the Mississippi River to the Pearl River—and annexed it to Louisiana. (Figure 79) This explains why Louisiana suddenly "jumps" across the Mississippi River and why it does so where it does.

Spain, which had its hands full with its increasingly rebellious colonies in Central and South America, finally agreed to negotiate its North American borders with the United States. In 1819, the Adams-Onis Treaty defined the western border of Louisiana as being the Sabine River from the Gulf of Mexico upstream to the 32nd parallel, at which point the boundary becomes a line due north to the Red River.

But why was the straight line located at the 32nd parallel? This location reveals the compromise reached by the United States and Spain over the hotly contested Red River. By locating the line at 32°, the United States acquired the southward flowing segment of the Red River, along with a protective buffer that, at its minimum point (in present-day Arkansas), is

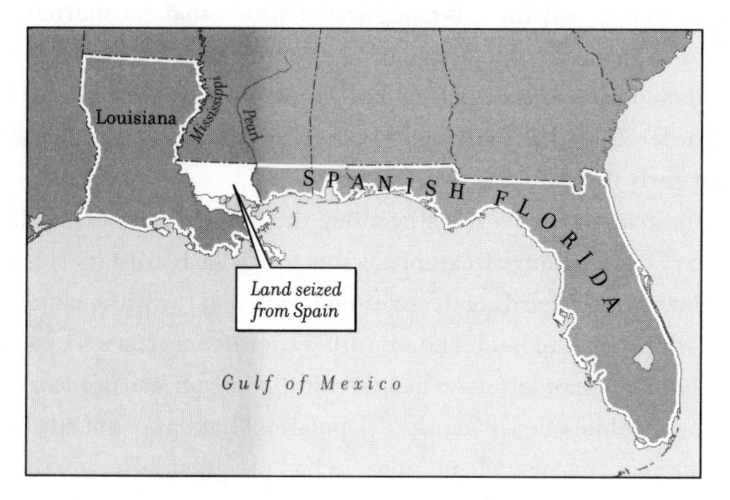

FIG. 79 The Area of Louisiana Seized from Spain—1810

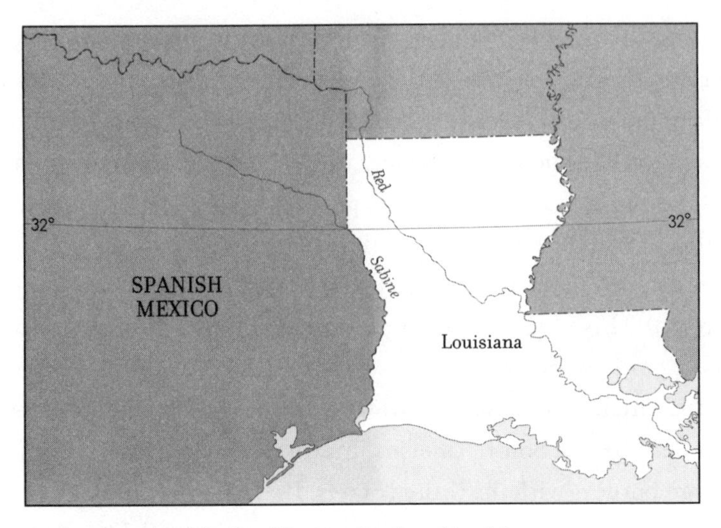

FIG. 80 Straight-line Western Border of Louisiana

exactly 10 miles wide. For its part, Spain got access to the southern bank of the eastwardly flowing upper segment of the river. (Figure 80)

Louisiana's Northern Border

Louisiana's northern border, also a straight line, is located along the 33rd parallel. But why there, particularly when the straight line that is its western border continues farther north? Was Louisiana shortchanged?

Louisiana wasn't shortchanged, since its northern border had been established fifteen years before the Adams-Onis Treaty defined its western border. Still, the location of Louisiana's northern border is linked to that early threat of war with Spain.

With Spanish troops massing along the Sabine, President Jefferson's secretary of war, Henry Dearborn, wrote to the local military commander for information regarding the size and location of the white population in the newly purchased land. The administration's concerns were partly military, but President Jefferson had the additional concern of incorporating into the republic a newly acquired population that was primarily French.

Some years earlier, when Congress had assigned Jefferson the task of recommending how to divide the Northwest Territory into future states, he shared his thoughts on the nature of state borders with James Madison. "Considering the American character in general," Jefferson wrote Madison in 1786, "a state of such extent as one hundred and sixty thousand square miles [roughly the size of California] would soon crumble into little ones." The reason such a large states would crumble, Jefferson feared, was that their vast size would embrace populations with conflicting interests. (Indeed, Jefferson's own home state of Virginia was a perfect illustration and eventually did divide into Virginia and West Virginia.)

One can see Jefferson's view in the northern border of Louisiana. At the time, there were a few areas of recent settlement by Americans on the Louisiana side of the Mississippi River opposite present-day Natchez and extending up to present-day Vicksburg. Farther down the Mississippi and in the bayous feeding the Gulf of Mexico, there were the French residents who constituted the bulk of the population. Newer French settlements extended up the Mississippi River to Baton Rouge and up the Red River. In addition, some Irish and German settlements had sprouted in the region of the Ouachita River. (Figure 81)

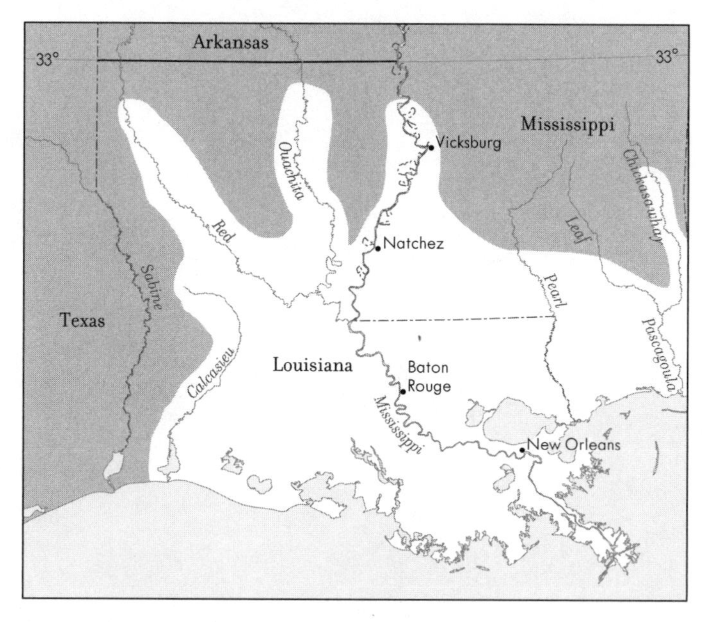

FIG. 81 Areas of Settlement at the Time of the Louisiana Purchase

Taken together, these settlements extended to about 32°30'. By locating the northern border of Louisiana at the 33rd parallel, Congress was acting in accord with Jefferson's philosophy on the boundaries of states. At 33°, Louisiana's northern border provided a bit of room for growth, while at the same time keeping the state small enough to secure the autonomy of the primarily French population. (A later Congress would demonstrate this philosophy again in locating the borders of another newly acquired non-English-speaking region, New Mexico.)

What is today an innocuous straight line at the top of Louisiana is, in fact, a line left behind by Thomas Jefferson that tells a tale of complex cultural and political events and a close call with war.

MAINE

How come Maine's boundary with Canada is where it is, when the St. Lawrence River seems like such an obvious boundary between the United States and Canada? Why are Maine's northern and eastern borders made up instead of lots of straight lines at different angles separated by squiggles? Why is Maine's western straight-line border located where it is, when New Hampshire clearly could have used a little more room?

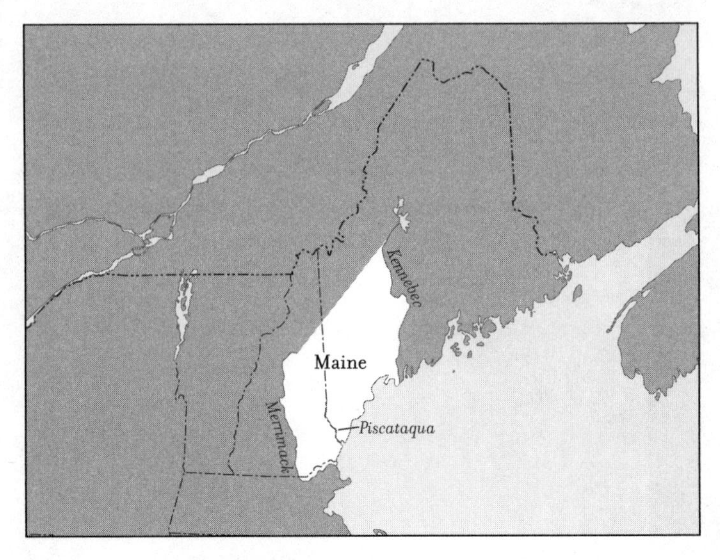

FIG. 82 The Province of Maine—1622

Europeans first claimed the area that is now Maine in the 1620 charter of the Plymouth Colony. This charter granted the colony all the land from the 40th parallel to the 48th parallel, between the Atlantic Ocean and the Pacific Ocean. This is a lot of land. It extends from what is today Philadelphia up past what is today the city of Quebec. Two years after its creation, the Plymouth Company Council deeded to two investors, Sir Fernando Gorges and Captain John Mason, the land between the Merrimack River and the Sagadahoc River (today known as the Kennebec). (Figure 82) The Council called this region the province of Maine.

Maine's Western and Eastern Borders

In 1629, Gorges and Mason divided their province along the Piscataqua River, which empties into the Atlantic at what is now the town of Portsmouth. As dividing lines go, the Piscataqua doesn't extend very far inland, but in 1629 Europeans didn't either, so the boundary served its purpose. Mason renamed his land, on the west side of the boundary, New Hampshire; Gorges named his land, on its east side, New Somersetshire. Gorges' name didn't last but the boundary did, for to this day the Piscataqua River remains the southwestern border between Maine and New Hampshire.

After the separation of New Hampshire, King Charles I issued a revised grant in 1639 further defining the borders of Maine. The Piscataqua River continued to serve as its southwest border, continuing up to the Salmon Falls River; the border then followed that river to its source. This section of the western border of Maine has continued to this day. Unlike today, the 1639 grant then stipulated a line to run northwesterly until reaching a point 120 miles from the mouth of the Piscataqua. Just how northwesterly wasn't specified and would, in the years ahead, become the subject of dispute.

Similarly, Maine's eastern border was the Kennebec River from the point where it empties into the sea to a point 120 miles upstream. Maine's northern border was then formed by connecting the two inland points of these eastern and western lines. (Figure 83)

Despite these boundary adjustments, Maine continued to be a realm under the jurisdiction of the Plymouth Colony, then, as the colony was later named, Massachusetts. Maine would, in fact, remain an extension

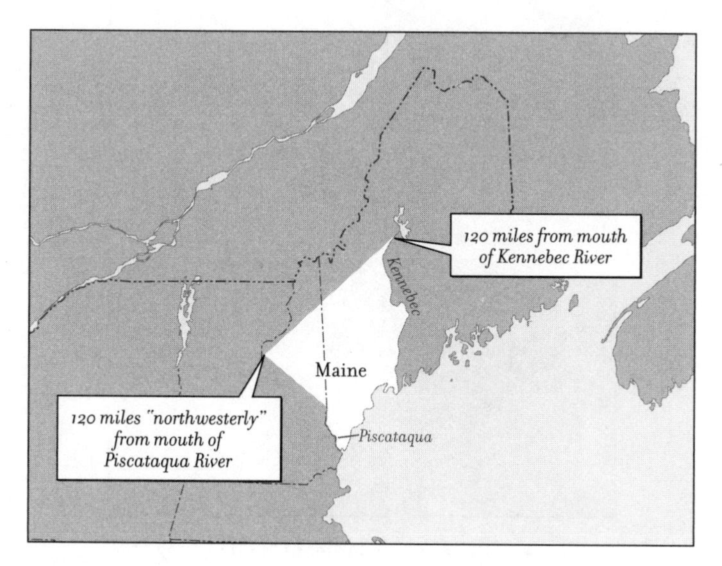

FIG. 83 **Maine—1639**

of Massachusetts until after the Revolution, which is why it is not counted among the original thirteen colonies.

Maine's feistier sibling, New Hampshire, had managed to separate itself from Massachusetts, in part because the British monarchy, still smarting from its recent (albeit temporary) overthrow by England's Puritans, lost no opportunity to cut Puritan Massachusetts down to size. Literally. One such opportunity involved New Hampshire's claims on its border with Maine. New Hampshire, over the years, had come to claim jurisdiction over increasingly more eastern land than Massachusetts thought was justified by the 1639 charter. In 1737, the King's Privy Council ruled in favor of New Hampshire's claim that its boundary with Maine was (as before) the Piscataqua River to the Salmon Falls River, but that now the line from the headwaters of the Salmon Falls River was to run "north two degrees west till 120 miles were finished." (Figure 84)

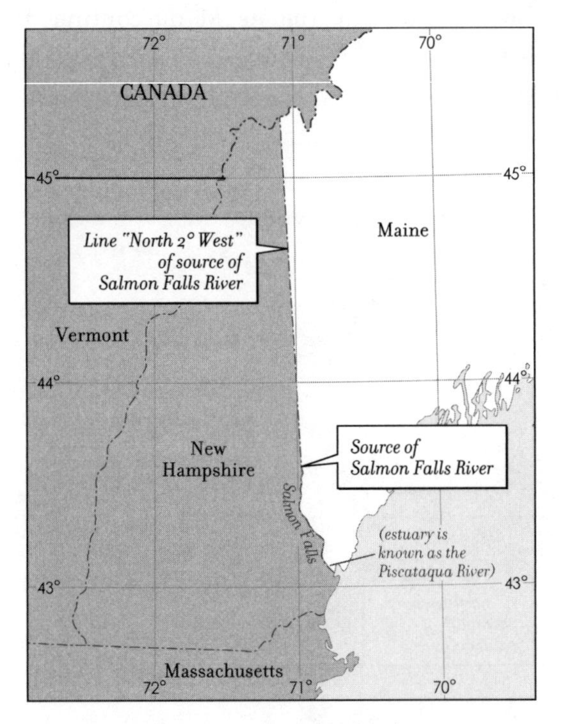

FIG. 84 The Western Border of Maine—1737 to Present

"North *two degrees* west"? Why not just north? The problem with "just north" is that the earth is not flat. If this boundary were truly due north, its line would appear curved on a flat map. To compensate, boundaries were often stipulated as some few number of degrees east or west of north. In Maine's case, the description seems to have sufficed, since this line has continued to serve as the border between Maine and New Hampshire to this day—the remains of a slap in the face to a state that is no longer even in the vicinity.

Maine's Northern Border

Maine's northern border was redefined in the 1783 Treaty of Paris, which ended the Revolution. In the wake of American independence, the two sides had to stipulate what constituted the United States. The northeast corner of the new nation, according to the treaty, was to be a line drawn due north from the source of the St. Croix River to the crest of the highlands that divide the rivers that flow to the St. Lawrence from those that flow to the Atlantic Ocean. This line was to continue westward until reaching the northwesternmost head of the Connecticut River. But this definition does not describe Maine's current northern border. A long dispute ensued over which highlands divide the rivers flowing to the St. Lawrence from those flowing to the Atlantic. (Figure 85)

Fifty years were to pass until this dispute was finally settled under the Webster-Ashburton Treaty. Why did it take so long? Couldn't they just float little boats to see which rivers went where? No, because all that would have shown was that the American interpretation was more accurate. And accuracy was not the issue. Montreal was. And the St. Lawrence River over which it presides.

The United States aspired to include Canada. It made no secret of this ambition. In the Articles of Confederation, Article XI states, "Canada acceding to this confederation, and adjoining in the measures of the United States, shall be admitted into, and entitled to all the advantages of this Union." In the War of 1812, American forces made several attempts

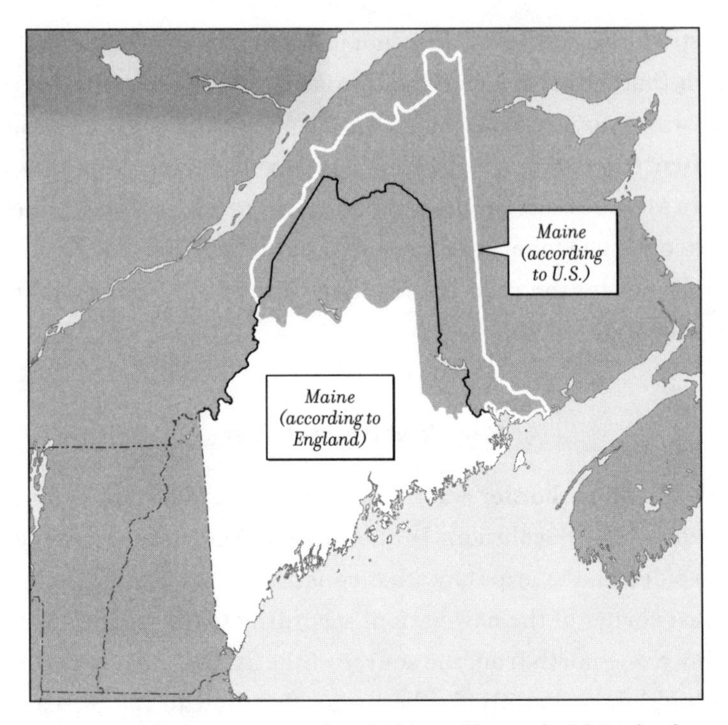

FIG. 85 The Northern Border of Maine—Disputed with England

to enter Canada in hopes of triggering rebellion from Great Britain. One of these attempts went through Maine.

An additional factor further delayed the negotiations. After becoming a state in 1820, Maine began granting land to settlers in the disputed regions. In response, Canada authorized its lumberjacks to take timber from these same regions. The diplomats, by this point, were arguing over such issues as, "What is a highland?" and "What is the Atlantic Ocean?" Maine, meanwhile, activated its militia. New Brunswick followed suit. Congress trumped New Brunswick by appropriating $10 million to equip fifty thousand soldiers.

At this point, Daniel Webster—not a man known to ask, "What is the Atlantic Ocean?"—took over the negotiations. In 1842, the northern border of Maine as we know it today acquired its final form in the Webster-Ashburton Treaty. The compromise boundary calmed British and Canadian fears by providing sufficient overland access to Quebec and Montreal. And Maine acquired access to the upper reaches of the St. John River.

Today, England and Canada are among America's closest allies. It is difficult to picture us as antagonists. The northern border of Maine, however, gives us another image, embedded in the American map.

MARYLAND

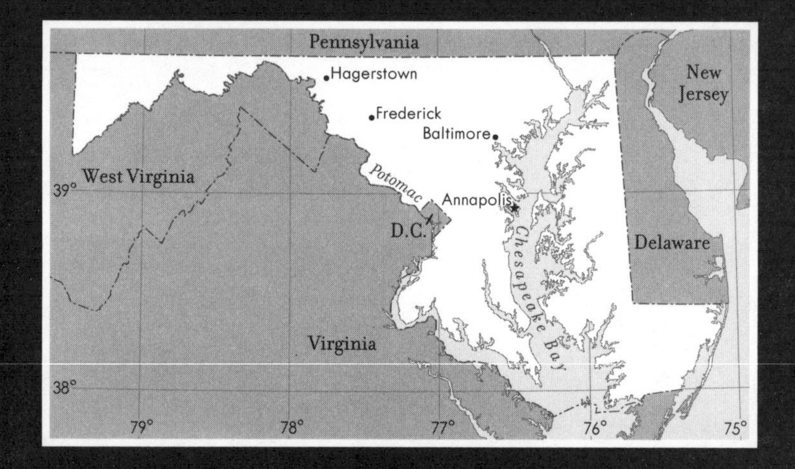

How did it happen that Maryland is almost broken in two? And that right-angled piece missing from its eastern edge—which is the state of Delaware—was Delaware really necessary? Who sliced off the southern end of Maryland's eastern edge? And why didn't they slice it straight? And why are Maryland's straight-line borders located where they are?

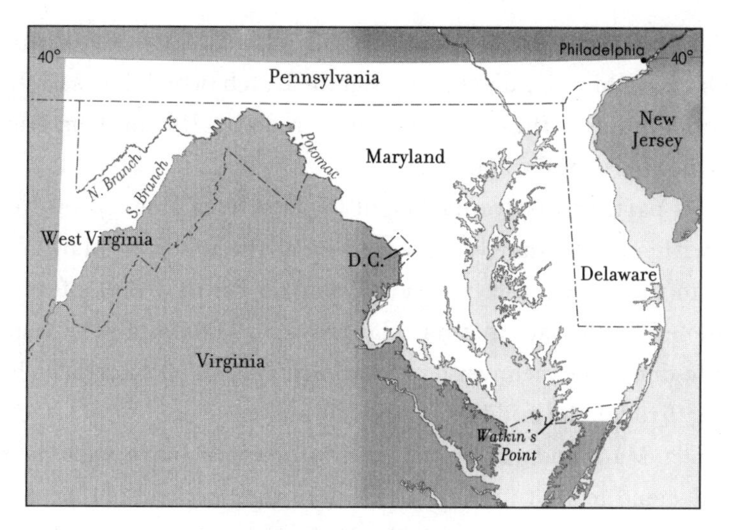

FIG. 86 Maryland (According to Maryland)—1632

Maryland was created by a royal charter issued in 1632 by King Charles I. The king, a Catholic, created Maryland to provide a place in the New World for England's Catholics. But this act, while full of good intentions, was also full of bad geography. Those errors led to a long history of border disputes between Maryland and every one of its neighbors. (Figure 86)

Maryland's Northern and Eastern Borders

According to its charter, Maryland's northern border was an east-west line located at 40° N latitude. Unfortunately, 40° N latitude turned out to be right in the middle of Philadelphia. Pennsylvania sought to have this line relocated, but its negotiations with Maryland became bogged down with another dispute between the two colonies, over, strange as it may seem, Delaware.

Delaware emerged some forty years after the creation of Maryland, when England finally succeeded in ousting the Dutch from North America. The southernmost realm of the Dutch had been the area around the Delaware Bay, which empties into the Atlantic Ocean at Cape Henlopen.

(See Figure 114, in NEW JERSEY.) This entire region was within the boundaries of Maryland's charter. But the Dutch (which is to say, Protestant) inhabitants of these settlements feared what life for them might be under the rule of Maryland's Catholics.

For its part, Pennsylvania sought to acquire this newly won region, since without it Pennsylvania would be at the mercy of Maryland for access to the sea via the Delaware Bay. And Pennsylvania's fears were heightened by the fact that its relations with Maryland were less than friendly, due to the inconvenient location of Maryland's northern border passing through the middle of Philadelphia.

In 1682, Maryland was denied possession of Delaware when the monarchy decided to rent it Pennsylvania. In the documents that were, in effect, a lease, Delaware was defined as consisting of the land within a 12-mile radius of the church at New Castle and all the land south of that circle as far as Cape Henlopen.

But Maryland did not believe that a lease agreement overrode a royal charter. Consequently, it took its case to the king, who promptly passed it to his Committee for Trade and Plantations. In 1685, the Committee ruled that Delaware was, in fact, a separate jurisdiction, since the area granted to Maryland was only intended to include land *uncultivated by Christians*. This may sound like a loophole to get the king off the hook, but, in fact, the second paragraph of Maryland's charter states that this land was being granted to start a colony "in a country hitherto uncultivated, in the parts of America, and partly occupied by Savages, having no knowledge of the Divine Being." Nasty words by today's standards, but it did the trick. Maryland's borders no longer encompassed Delaware.

"For avoiding further difference," the Committee proposed a formula for equally dividing the area between the ocean and the Chesapeake Bay, north of the latitude of Cape Henlopen. Unfortunately, this formula, like the charter that preceded it, was filled with errors. It wrongly assumed that the Chesapeake Bay extended to the 40th parallel, and its parameters resulted in a dividing line that sliced into the 12-mile radius at Delaware's northern end. (See Figure 47 and more details in DELAWARE.)

For the next century, Maryland's eastern border would remain disputed ground. Amid what appeared to be a stalemate, however, certain elements of what ultimately became Maryland's eastern and northern border did begin to emerge. In 1732, Maryland's colonial governor, Lord Baltimore, was in London, yet again negotiating the colony's boundaries. This time, however, he found his adversaries surprisingly willing to negotiate. Pennsylvania, representing Delaware, accepted Lord Baltimore's proposed eastern border, and in return, Lord Baltimore agreed to Pennsylvania's proposal that Maryland's northern border be relocated 15 miles south of South Street in Philadelphia. Only later did Lord Baltimore learn that the reason he'd succeeded was that his map *mistakenly located Cape Henlopen!* On his erroneous map, Fenwick Island, nearly 25 miles to the south, was identified as Cape Henlopen. (Figure 87)

Upon discovering the error, Lord Baltimore demanded that the border be negotiated again. Pennsylvania said no. The short-term effect was

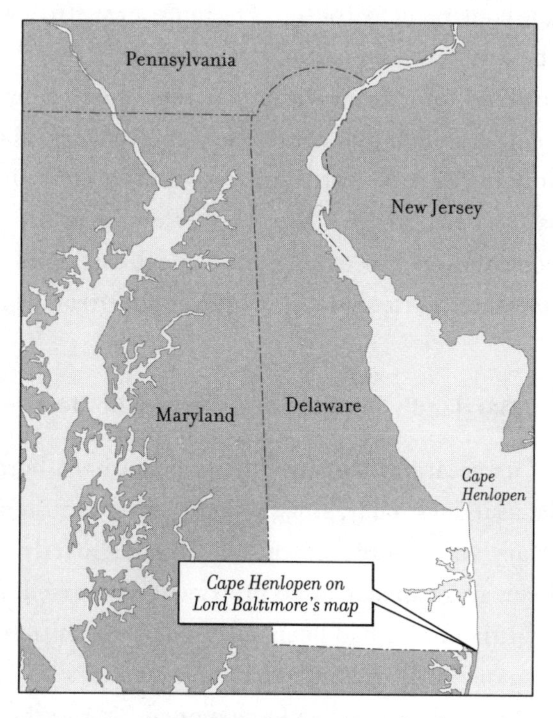

FIG. 87 The Eastern Border of Maryland—Impact of a Bad Map

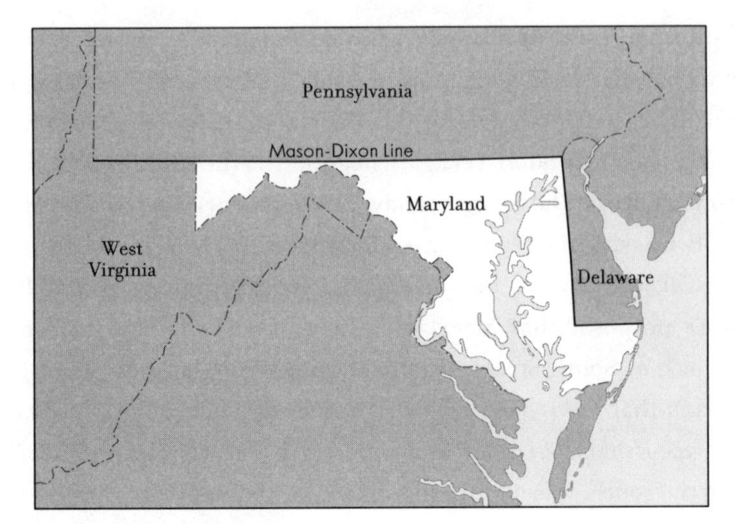

FIG. 88 The Mason-Dixon Line—Border of Four States

that the dispute continued to drag on. The long-term effect was that Fenwick Island became, and remains, the southern border of Delaware. Likewise, Maryland's northern border has remained a line 15 miles below South Street, Philadelphia. For the next thirty years, Maryland protested what it believed to be two manifestly unfair consequences of an erroneous map. Finally, in 1763, Maryland relented, joining with Pennsylvania to commission two of England's most esteemed scientists to survey their border: Charles Mason and Jeremiah Dixon. (Figure 88)

Maryland's Southern and Western Borders

At the time of its creation, Maryland had no southern border opposite West Virginia, as it does today, since West Virginia was then part of Virginia. And Maryland's border with Virginia, primarily the Potomac River, would certainly appear to be pretty cut-and-dried. And yet, the Maryland/Virginia border had been a source of contention dating back to *before* the issuance of Maryland's charter!

The original boundary Charles I envisioned for Maryland would not have chopped off the land that stretches southward between the Chesapeake Bay and the Atlantic Ocean. When Virginia learned of the king's intent, it protested. Members of the Virginia Colony had already migrated from its foundations along the James River across the Chesapeake, where they'd established plantations. These lands, after all, were originally part of the Virginia Colony. But if these Virginians (which is to say, Protestants) were now to be within the jurisdiction of Maryland (which is to say, Catholics), what sort of treatment could they expect? Rather than foster discord, Charles I amended his plan, calling for the boundary to divide the southern portion of the peninsula with a line from Watkin's Point due east to the ocean.

But the line the king described in Maryland's charter is not the line that exists today, because once again of a number of errors. Watkin's Point, for starters, no longer existed, having already eroded away. The local men chosen to survey the line in 1666 attempted to figure where Watkin's Point had been, but, it was later discovered, they figured wrong. In addition, they veered to the north as they marked off what was supposed to be a line due east. (Figure 89)

Maryland demanded that the line be redone. But its protest was drowned out by louder protests that dominated the era, regarding the need to unite to fight the French and Indian War. Afterward, Maryland

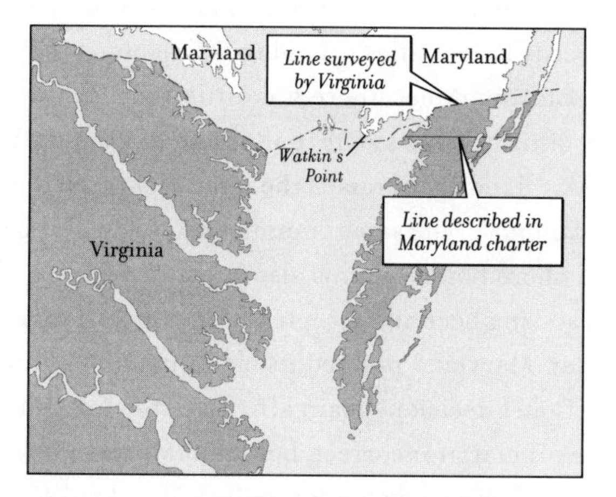

FIG. 89 Maryland's Southern Border on the Eastern Shore Peninsula

protested again but was again drowned out, this time by cries for unity to fight the Revolution, then to create the United States, then to fight the War of 1812 . . . Not until 1877 did the federal government appoint a special commission to consider the matter. By this time, the region was populated with farms whose property lines were aligned with their respective state lines. Consequently the commission ruled that the erroneous line should stand, since it had functioned for so many years.

The Potomac River, too, has caused more conflict than one might expect. In this case, the problem stemmed from the fact that, at the time Maryland's charter was issued, no one knew exactly where the western reaches of the Potomac were. One near disaster regarding this border was only narrowly (*literally* narrowly) averted when the mishap-prone Lord Baltimore agreed to a northern border 15 miles south of Philadelphia. The point in western Maryland where the state almost breaks in two is the result of the Potomac flowing in a northerly arc that nearly touches that relocated border.

Farther to the west, the Potomac divides into northern and southern branches. In marking its boundary, Maryland followed the South Branch of the Potomac, since it is the larger of the two branches. But Virginia claimed that the North Branch of the Potomac was the intended southern boundary of Maryland. (Figure 90)

Maryland protested Virginia's claim, but Virginia, being the older colony, had already issued titles to land in the disputed region. Since the king opted not to intervene, there was little that Maryland could do. When West Virginia became a state in 1863 and its boundaries needed to be approved, Maryland again raised the issue regarding the appropriate branch of the Potomac. The same commission that ruled against Maryland's Eastern Shore border claims also ruled that the North Branch of the Potomac had long been the accepted border and therefore would remain the border. Maryland pressed its claim further, culminating in a 1910 Supreme Court decision, again affirming the North Branch of the Potomac as the official, if incorrect, border of Maryland. With the resolution of Maryland's southern border in 1910, Maryland's *western* border was then officially established, that being a line due north from the

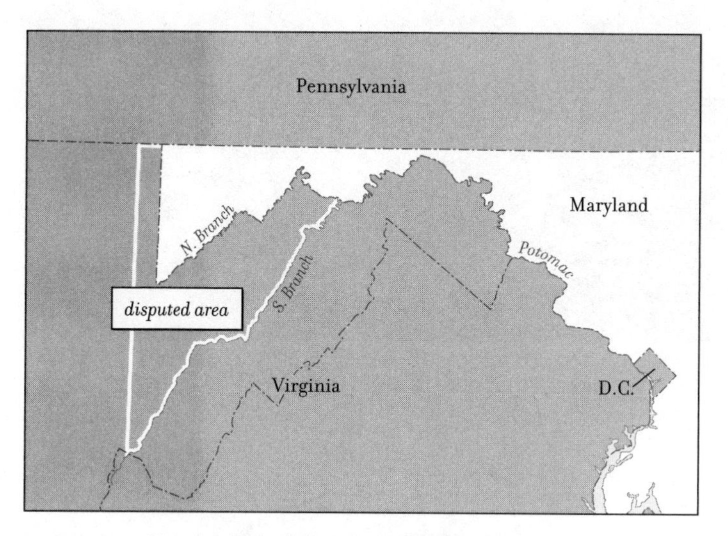

FIG. 90 **Maryland/Virginia—Western Boundary Dispute**

headwaters the Potomac (northern branch) to its boundary with Pennsylvania.

Each of the borders stipulated in Maryland's 1632 charter turned out to have been in error. One might say that Maryland is the shape of human error. But the irregularities of its border also contain another important fact. In the wake of so many mistakes and defeats, Maryland has survived and even thrived.

MASSACHUSETTS

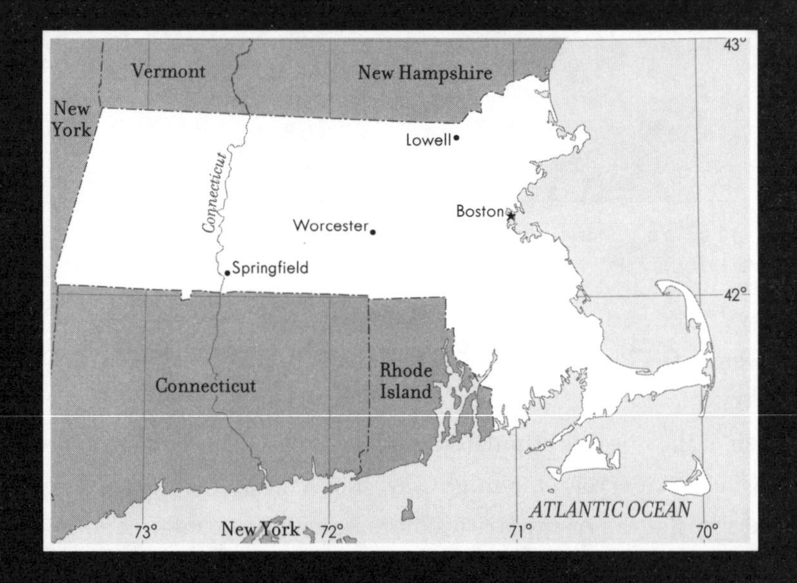

How come Massachusetts starts out nice and wide on the coast and then suddenly gets sliced on its top and bottom? Why is its western border angled instead of due north? And what's with the tiny snip taken off the southwest corner of Massachusetts?

What we know today as Massachusetts is actually an amalgam of two colonies, the Plymouth Colony and the Massachusetts Bay Colony. The first of these colonies was the Plymouth Colony. Its royal charter, issued by King James I in 1620, granted it all the land between 40° and 48° N latitude, from the Atlantic Ocean to the Pacific Ocean. If these were still its borders, the Boston Red Sox, New York Yankees, New York Mets, Pittsburgh Pirates, Cleveland Indians, Detroit Tigers, Chicago White Sox, Chicago Cubs, Milwaukee Brewers, Minnesota Twins, Toronto Blue Jays, and Seattle Mariners would all be playing for Plymouth.

Massachusetts' Southern Border

The first settlement of the Plymouth Colony was established in, not surprisingly, the town they named Plymouth, in what is now the southeast corner of Massachusetts. It is the settlements of the Plymouth Colony that account for the fact that this southeast region remained affixed to Massachusetts, even when Rhode Island and Connecticut were carved away.

Nine years after establishing the Plymouth Colony, a new group came over from England and settled to the north, at Boston and along the Massachusetts Bay. King Charles I issued this Massachusetts Bay Colony a separate charter in 1629. That charter established what led to the straight-line southern border of Massachusetts when it stated that the southern border of the Massachusetts Bay Colony shall be 3 English miles to the south of the southernmost point of Massachusetts Bay.

But this location is not the southern border of the Massachusetts we now know. Rather, its southern border is *based* on this description. Massachusetts did not survey its southern border until a number of its colonists had created settlements at Hartford, New Haven, and New London. In 1639, these settlements confederated under the Fundamental Orders of Connecticut. Three years later, as the Massachusetts Bay Colony merged with the Plymouth Colony to form Massachusetts, the consolidated colony agreed to a border being created that separated Massachusetts and Connecticut. But the line drawn by Massachusetts was, according to Connecticut, 8 miles too far south. Since Massachusetts was now an amalgam of two colonies, how did Connecticut come up with this figure?

Connecticut based its claim on the charter of the Massachusetts Bay Colony, though (perhaps out of deference to the location of the Plymouth Colony) it applied a liberal interpretation to its language. Where the charter described a southern border as being 3 miles south of the southernmost point of Massachusetts Bay, Connecticut allowed the words to mean 3 miles south of the southernmost point of the waterways feeding into Massachusetts Bay. The tributary to Massachusetts Bay that extends farthest to the south is the Neponset River, and 3 miles south of its southernmost point is where Connecticut believed its border with Massachusetts should be located. (See Figure 37, in CONNECTICUT.)

Numerous surveys followed, none of which were acceptable to the side that had not hired the surveyor. Adding to the friction, the previously established towns of Enfield, Somers, Suffield, and Woodstock, located in the disputed zone, declared their intention to consider themselves to be within the jurisdiction of Connecticut. (See Figure 38, in CONNECTICUT.)

After the Revolutionary War, the two states agreed to a Boundary Commission. The border that resulted, and that remains to this day, was a straight east-west line 3 miles south of the southernmost point of the Neponset River—but it has a sizable dip into Connecticut toward its western end and has a small blip into Connecticut farther east. The blip is where the Connecticut River crosses into Connecticut. In this era that preceded highways and railways, rivers were vital economic arteries. The blip in the border results from the boundary following the crest of the hills abutting the Connecticut River in the area where it crosses the border.

The border's jog farther west into Connecticut was compensation to Massachusetts for the settlements it was losing with this line. Massachusetts received five-eighths of the land around Congamond Lakes. (Figure 91) Known as the Southwick Jog, it remains to this day in the Massachusetts border with Connecticut.

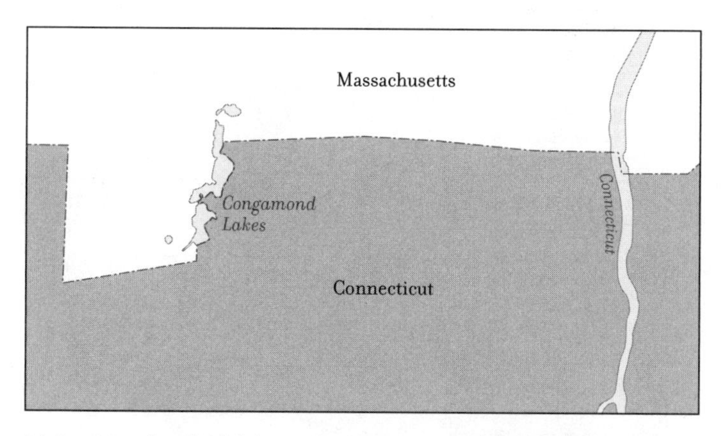

FIG. 91 **Southwick Jog—1804 Agreement with Connecticut**

Massachusetts' Northern Border

The second border of Massachusetts to get sliced off from the colonial charters was its northern border. In this case, the loss of the land came in two stages. Two years after the founding of the Plymouth Colony in 1620, the king deeded to two investors, Sir Fernando Gorges and Captain John Mason, the land between the Merrimack and the Kennebec rivers. Mason and Gorges ended their partnership in 1629, dividing their land by using the Piscataqua River as the border. Mason called his land, to the west of the Piscataqua, New Hampshire; Gorges called his land, east of the Piscataqua, New Somersetshire (but everyone else called it Maine). The entire investment, as it turned out, was a bust, and in 1651, Massachusetts reasserted its claim to these lands. While Maine continued as a part of Massachusetts until 1820, New Hampshire, being more populated than Maine, resisted.

In 1680, King Charles II issued New Hampshire a charter for self-government. Since the charter did not specify the boundaries of New Hampshire, Massachusetts assumed its border with New Hampshire was that stated in the charter of the Massachusetts Bay Colony. Accordingly, Massachusetts assumed its northern border was located along a line 3 miles northward and eastward of the Merrimack River. New Hampshire disagreed. It argued for a line beginning 3 miles north of

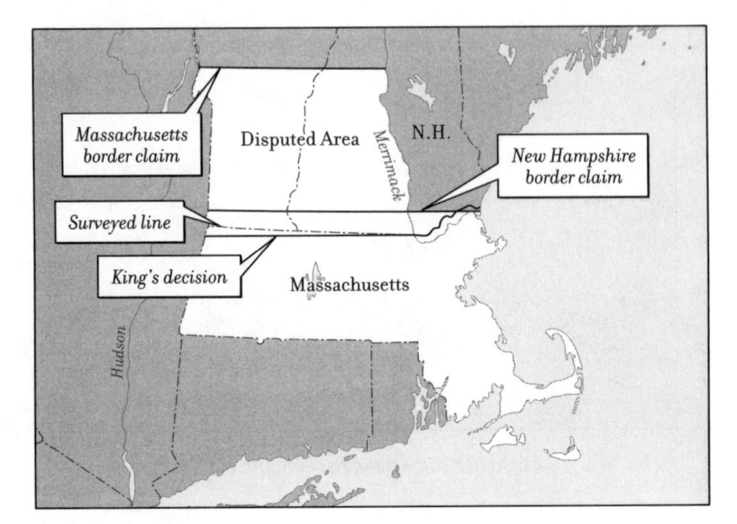

FIG. 92 **Massachusetts/New Hampshire Border Dispute**

the mouth of the Merrimack, as specified in the charter, but then simply heading due west to the New York border. The dispute continued until 1741, when King George II drew his own line. It was even more favorable to New Hampshire than the one New Hampshire had proposed! (Figure 92)

The king's line began 3 miles north of the mouth of the Merrimack and, as stated in the original charter, paralleled the river, but upon reaching the point where the Merrimack turns north, a straight line due west took over. Despite being the big winner in this dispute, New Hampshire didn't hesitate to point out (correctly) that the line, as surveyed, veered to the north of the line described by the king. But New Hampshire's complaint was lost in the accelerating winds of the Revolution, and the inaccurately surveyed line remains in place to this day as the northern border of Massachusetts.

But why would the king so favor New Hampshire over Massachusetts? There were two reasons. First, the king saw Massachusetts, a Puritan colony, as an extension of the British Puritans who had recently (though only temporarily) overthrown the monarchy. Second, New Hampshire had undertaken a policy of lavishly praising the monarchy and declaring its commitment to the Anglican Church.

Massachusetts' Western Border

The western border of both the Plymouth Colony and Massachusetts Bay Colony, according to their charters, was the Pacific Ocean. Blocking the way, however, were the Dutch, who had their own charters granting them the land between the Connecticut River and the Delaware River. (See Figure 114, in NEW JERSEY.) All this land came under British rule in 1674, when the British ousted the Dutch from North America. By defeating the Dutch, England had eliminated one major headache . . . but acquired another. The boundaries of its newly acquired colony, which the British named New York, were those inherited from the Dutch—and they conflicted with, among others, those of Massachusetts. Massachusetts argued that its western border should be the Hudson River (thus giving the western part of the colony access to the sea). New York claimed its border with Massachusetts should be the Connecticut River, as the Dutch had claimed. (See Figure 121, in NEW YORK.)

Unable to resolve their dispute, the two colonies submitted the issue to England. In 1759, the British government declared that the boundary between New York and Massachusetts would be a straight line 20 miles

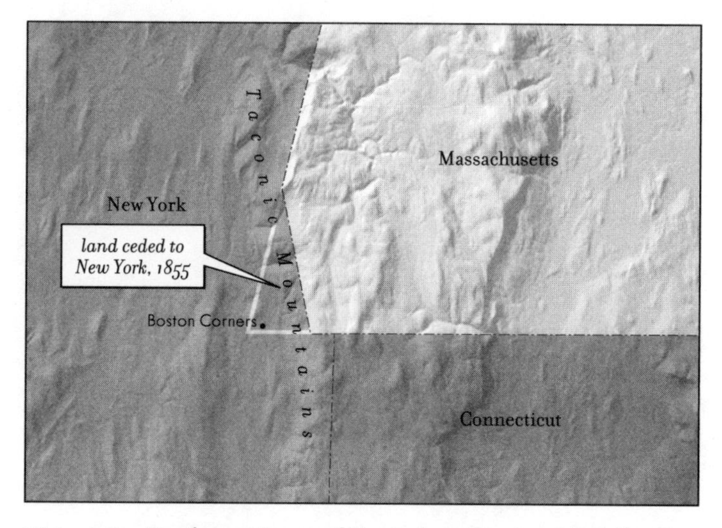

FIG. 93 Southwest Corner of Massachusetts—Land Ceded to New York

east of the Hudson River. That line remains the western border of Massachusetts to this day.

With one exception. The exception is a little snip off the southwest corner of Massachusetts. (Figure 93) But this little snip indicates how far we had come in terms of cooperation between jurisdictions. The area at issue was the town of Boston Corners.

Boston Corners may look innocent enough on a map, but it happened to be located in the mountains in such a way that, at the time, it was accessible by road only through New York or Connecticut. Since Massachusetts authorities had no access, Boston Corners had become a haven for the less than law-abiding. For this reason, Massachusetts suggested giving this corner to New York in 1853. New York accepted the offer, equally desirous of cleaning up the area. In 1855, Congress recognized this cooperative act by approving the transfer of land.

How come Michigan has that whole separate section that's actually attached to Wisconsin? And why is Michigan's southern border made up of two straight lines instead of one?

Before Michigan was Michigan, it was part of the Northwest Territory. The proposed boundaries for the state-not-yet-named Michigan were considerably simpler than those that exist today. Michigan's southern border was envisioned in the Northwest Ordinance (1787) as a line due

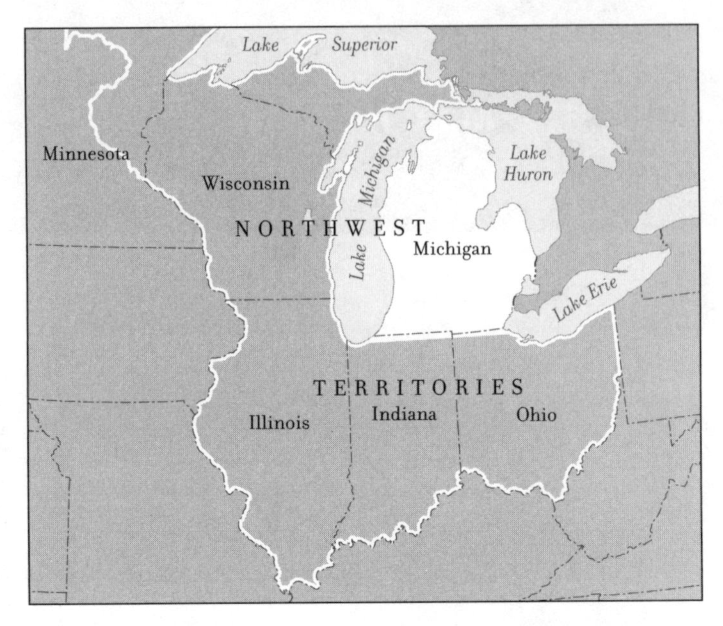

FIG. 94 Michigan According to the Northwest Ordinance

east from the southernmost point of Lake Michigan to its intersection with Lake Erie. Everything to the north was Michigan up to the Canadian border. (Figure 94)

Michigan's Eastern Border

The oldest of Michigan's existing borders is its eastern border. This boundary was established in the 1783 Treaty of Paris, which officially ended the American Revolution and defined the borders of the new United States. The section of the treaty separating what is today Michigan from what is now Canada specified that the boundary be located through the middle of Lake Erie

> until it arrives at the water communication between that lake and Lake Huron; thence along the middle of said water communication into Lake Huron, thence through the middle of said lake to the water communication between that lake and Lake Superior.

By locating the boundary along the middle of these lakes and the waterways connecting them, the two nations were preserving vital avenues of transportation for both sides.

Michigan's Southern Border

When settlers began pouring into what is now Ohio and southern Michigan in the early 19th century, Congress began to subdivide the Northwest Territory into the territories that would later become states. In doing so, it discovered that a line due east from the southernmost point of Lake Michigan to Lake Erie, as stiputed in the Northwest Ordinance, would cut off Ohio from Toledo, its valuable western port. Ohio got Congress to redefine its border with Michigan such that its eastern end was located on Lake Erie just above Toledo. From this point, the line proceeded on a direct course aimed at the southernmost point of Lake Michigan, though it stopped at the Ohio/Indiana border. This correction accounts for the eastern half of Michigan's southern border being slightly angled. (See Figure 136, in OHIO.)

Michigan was less than pleased. But lacking the population that Ohio had, and still needing congressional approval for its own statehood, there wasn't much it could do. Then Indiana got into the act.

When the territory of Indiana came into existence, it discovered that it, too, had a problem regarding access to the Great Lakes. If, as stipulated in the Northwest Ordinance, Indiana's northern border was to be a line that is tangent with the southernmost point of Lake Michigan, Indiana would have only one infinitely small point of access to the lake. With the rapidly increasing possibility of a canal (what would, in time, be the Erie Canal) connecting the Great Lakes to the Hudson and the sea, access to the Great Lakes acquired enormous economic significance. Consequently, in 1816, Congress relocated Indiana's northern border 10 miles north of the southernmost point of Lake Michigan. From this new location the Indiana/Michigan border extended due east to the longitude of the Indiana/Ohio border. Congress, in short, lopped off another strip of Michigan. As a result, Michigan's southern border consists of two straight lines that are slightly offset. (See Figure 136, in OHIO.)

But Michigan didn't want two straight lines that are slightly offset. It wanted Toledo and Gary—important ports on Lakes Erie and Michigan. So passionate were Michigan's feelings that in 1835 it dispatched its territorial militia to take the disputed land by force. The ensuing conflict, known as the Toledo War, resulted in the capture of nine surveyors working for Ohio and the stabbing of a Michigan sheriff (though there is some dispute as to whether the stabbing was part of the war or a tavern brawl).

Michigan's Upper Peninsula

To end the violence, Congress offered Michigan compensation for the taken land, in the form of a large peninsula Congress took from what was to be Wisconsin. Michigan took the deal.

Wisconsin, on the other hand, took offense. But it lacked the political clout to prevent it—just as Michigan had lacked the clout to fend off Ohio and Indiana. Years later, however, Wisconsin did win a dispute with Michigan over the location of the Upper Peninsula's boundary line. (For more on this, go to WISCONSIN.)

There is a sense in which Michigan's irregular and, in the case of its Upper Peninsula, seemingly inexplicable borders are a reflection of ourselves. Michigan's theoretical borders were a beauty to behold: three Great Lakes connected by waterways, resting on a straight line base. The border that ultimately emerged still reflects this vision, but it is bent and blemished by mundane, albeit vital, needs.

MINNESOTA

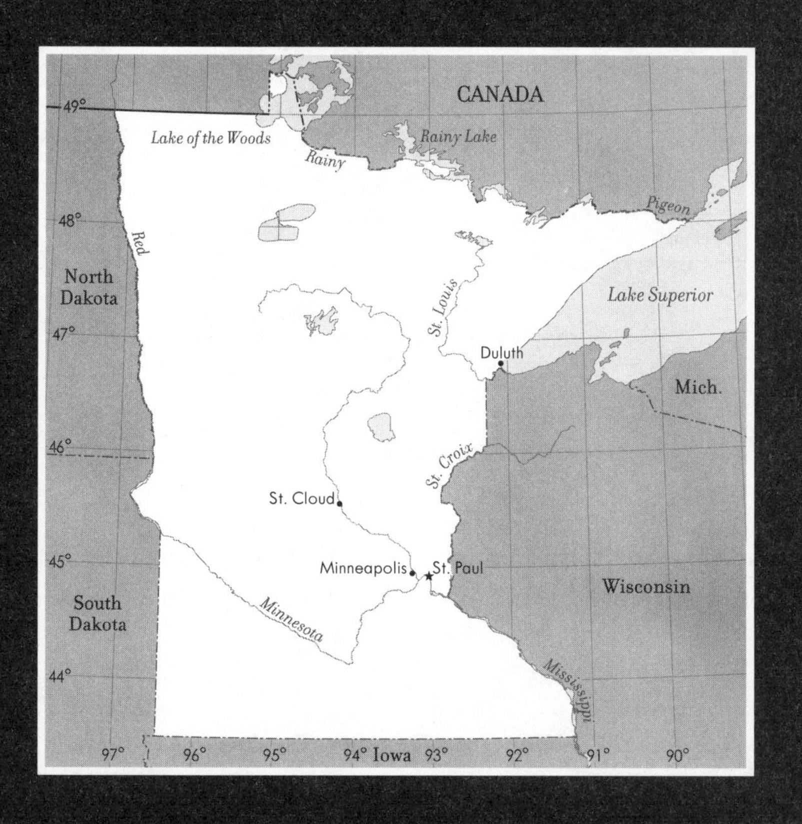

What in the world is that little blip up on top of Minnesota? Why isn't the straight-line part of Minnesota's northern border located at the top of the blip? How come Minnesota's southwest corner just misses lining up with the corner of its neighbor underneath, Iowa?

The area that is today called Minnesota was acquired by the United States in two separate events. The section of Minnesota that is east of the Mississippi River was part of the Northwest Territory, the region

FIG. 95 **Minnesota's Antecedents**

England and its American colonists won from France in the French and Indian War (1754–63). The section west of the Mississippi was acquired in the Louisiana Purchase (1803). (Figure 95)

Minnesota's Northern Border

The first of the boundaries of present-day Minnesota to surface was its northern border. After the American Revolution, the land acquired in the French and Indian War became part of the United States (where it became known as the Northwest Territory). Under the 1783 Treaty of Paris, which ended the war, the borders of the new United States included a line that threaded its way from the western end of Lake Superior through the network of lakes leading to the northwest corner of Lake of the Woods. "And from thence," the treaty said, "on a due west course to the river Mississippi."

Here, then, the northern border of Minnesota began to emerge. But there was a problem. The Mississippi doesn't extend as far north as Lake of the Woods. The United States and England remedied this error in a later treaty that delineated the border between the Louisiana Purchase and what was then called British North America (Canada). This time, the boundary at the northwest point of Lake of the Woods was described as a line due north or south to the 49th parallel and then due west along that parallel to the Rocky Mountains. (To find out why they chose the 49th parallel, go to DON'T SKIP THIS.) As it turned out, the 49th parallel was south of the northwestern corner of Lake of the Woods. This explains how Minnesota got that little blip on its top.

Minnesota's Eastern Border

Minnesota inherited its eastern border from Wisconsin. When Wisconsin's borders were finalized for statehood, Congress was already contemplating its future neighbor to the west, and adjusted Wisconsin's boundary to provide for a more equitable division of resources. Originally, Wisconsin's western border extended to the Mississippi River, since the French and Indian War wrested from France all the land between the Ohio and Mississippi rivers. Had Wisconsin retained this boundary, Minnesota would have had no frontage on any of the Great Lakes. So Congress altered Wisconsin's western border, having it depart the Mississippi at its juncture with the St. Croix River. The Wisconsin/Minnesota border then follows the St. Croix to a point due south of the westernmost point of Lake Superior. A straight line then connects these points. (Figure 96) As a result, Wisconsin fronts the southern shore of Lake Superior and Minnesota fronts the northern shore. These boundary choices vividly reflect the nation's commitment to the principle that all states should be created equal.

Minnesota's Southern Border

As with its northern and eastern borders, Minnesota also inherited its southern border, which had previously been established as Iowa's northern border. As with Wisconsin, Congress had created Iowa with its future

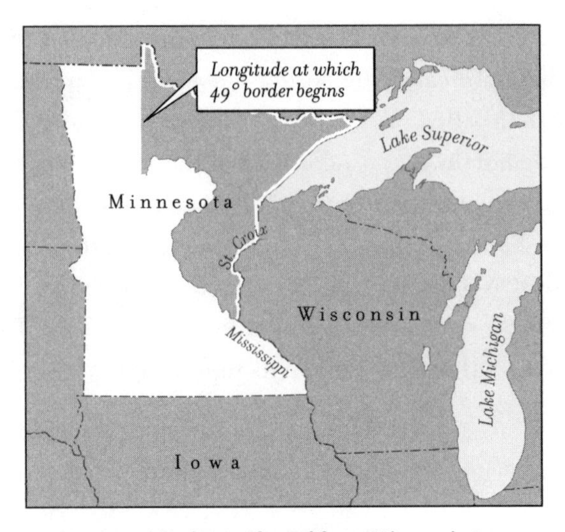

FIG. 96 Land Transferred from Wisconsin to Minnesota

neighbor to the north in mind. It rejected a proposal from Iowa that would have placed the city of St. Paul within its borders. And the border Congress imposed, an east-west line located at 43°30', preserved what was then called the St. Peter's River (now known as the Minnesota River) and its tributaries for a region that otherwise had lots of lakes that went nowhere. (Figure 97)

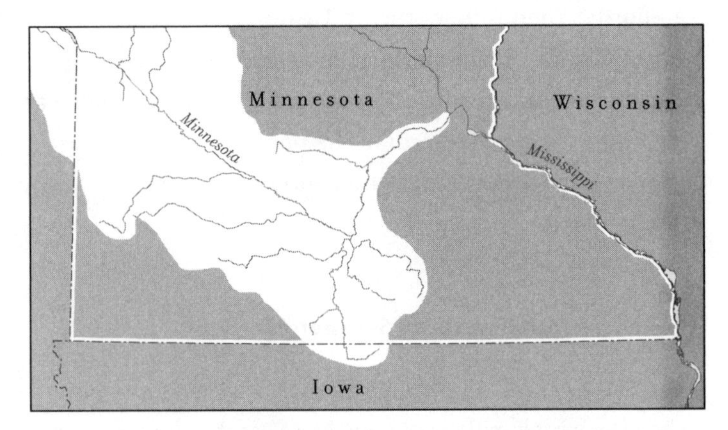

FIG. 97 Minnesota's Southern Border—Watershed of Minnesota River

Minnesota's Western Border

When Minnesota became a state in 1858, its western border was defined for the first time. Congress opted for a series of waterways running virtually due north. At the northwest corner of the state, the Red River of the North provides the boundary for nearly 200 miles before its path veers to the east. The point where it turns, however, is its juncture with the Bois de Sioux River, which takes over as the western border up to its source at Traverse Lake, which then provides an ideal north-south line until turning the task over to Big Stone Lake. Only at the southern end of

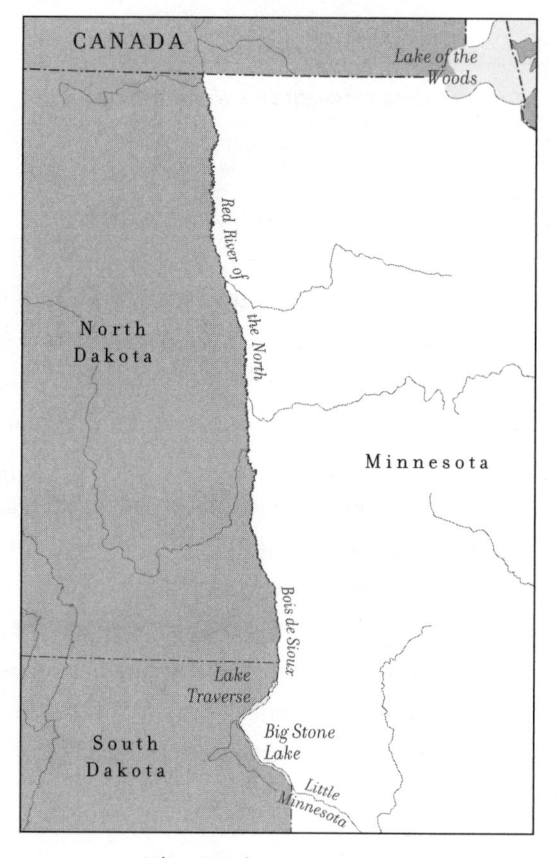

FIG. 98 Minnesota's Western Border—Connection of Rivers and Lakes

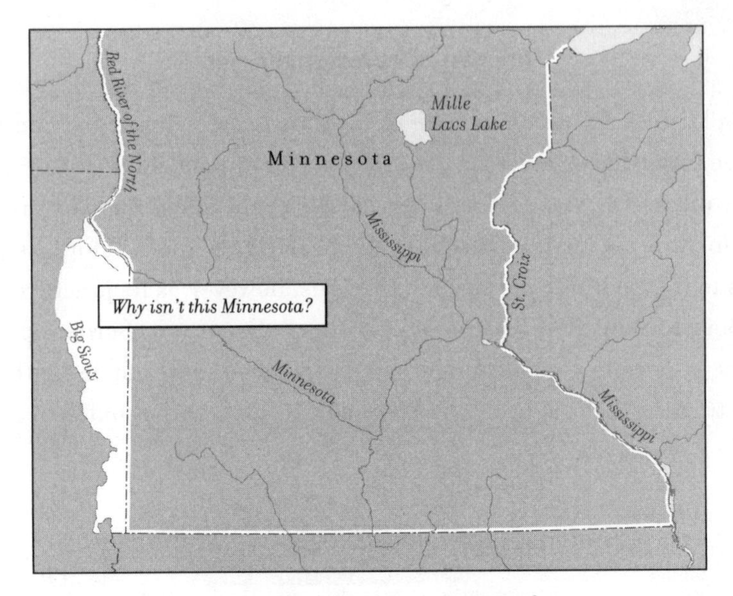

FIG. 99 Minnesota's Straight-Line Western Border

Big Stone Lake is a straight line introduced, which runs 130 miles due south to the Iowa border. (Figure 98)

But this straight-line element in Minnesota's western border does not line up with the corner of Iowa. What's more, yet another river is available that, in fact, forms the upper segment of Iowa's western border: the Big Sioux River. Why did Congress opt not to use the Big Sioux to complete Minnesota's western edge and align it with that of Iowa? (Figure 99)

Indeed, Minnesota's statehood delegation proposed exactly that border. But the region had not yet been officially surveyed. Minnesota delegate Henry M. Rice urged the Government Land Office to do so, but the Land Office wouldn't comply because such a survey was not in its budget.

In a sense, the southwest corner of Minnesota is a very special place. True, it fails to line up with the corner of Iowa, but this misfit with Iowa remains today as a monument to something we tend not to associate with the good old days of America's pioneering past: government bureaucracy.

MISSISSIPPI

Why does Mississippi have that little tab at the bottom of the state? And how did it lose that little snip at the top of the state? Why is Mississippi's northern border on the same line as Alabama's and Geor-

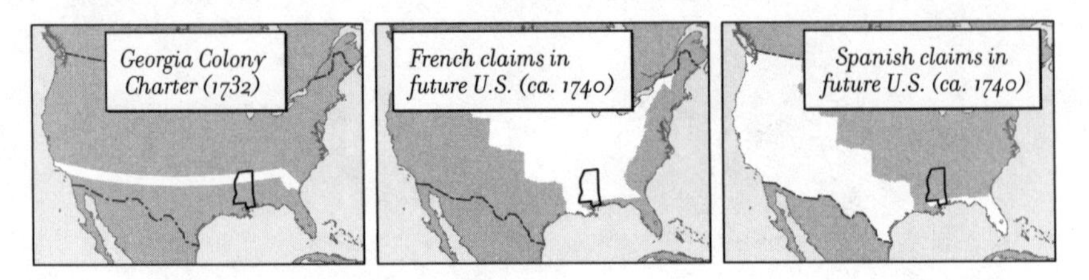

FIG. 100 Mississippi's Predecessors

what is now Mississippi and parts of what is now Georgia. And Spain claimed the land that's now the southern end of Georgia, Alabama, and Mississippi. (Figure 100)

After the Revolution, Georgia, along with the other colonies with vast land claims, released its western land so that new states could be formed. Its action was not without self-interest. The conflict over slavery served as a strong inducement to create as many pro-slavery (or in the north, pro-abolition) states as possible, so each side could have enough votes in the U.S. Senate to maintain (or, for the North, abolish) slavery.

Congress affixed the initial boundaries of the Mississippi Territory as the Mississippi River on the west (which eventually became the western border of the state of Mississippi); the Chattahoochee River on the east; on the north, a line to be drawn due east from the mouth of the Yazoo River to the Chattahoochee River; and on the south, 31° N latitude. (Figure 101)

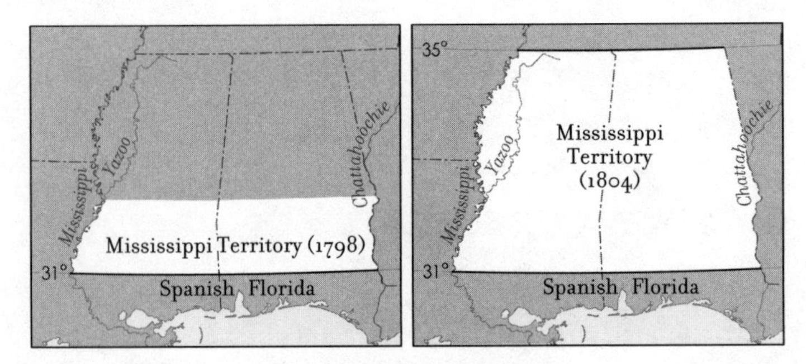

FIG. 101 Expansion of Mississippi Territory

Mississippi's Northern and Southern Borders

Congress was treading carefully in the creation of the Mississippi Territory. France, which had been our ally in the Revolutionary War and which remained a vitally needed friend, had fur trappers operating in what is today the northwest region of Mississippi. And Spanish claims to Florida had been established by treaty at the 31st parallel. (For the details of Spain's claims, go to FLORIDA.) Still, the United States wished to assert its claim, since failure to do so could have risked losing the land by default.

With the Louisiana Purchase (1803), concerns about conflicts with France evaporated. Consequently, one year after the Louisiana Purchase, Congress amended the northern border of the Mississippi Territory, having it correspond to the northern border of its parent, Georgia. This is why Mississippi, Alabama, and Georgia all share the same northern border to this day.

There was only one remaining problem with the Mississippi Territory. Nowhere along its southern border did it have direct access to the Gulf of Mexico. Spain held the land beneath the Mississippi Territory. It was known as West Florida and extended all the way along the Gulf of Mexico to the Mississippi River.

But Spain was becoming a doddering empire, not nearly so powerful as it had once been. In 1810, the United States seized the westernmost end of Florida, a chunk that extended from the Mississippi River to the Pearl River. This land was annexed to Louisiana. Three years later, American forces seized the adjacent chunk of land, extending from the Pearl River to the Perdido River. This parcel was annexed to the Mississippi Territory. (Figure 102)

Mississippi's Eastern Border

Congress often created territories knowing their borders would eventually be too large for states. But until a territory acquired sufficient population, one big jurisdiction was often more efficient. In 1816, to enable

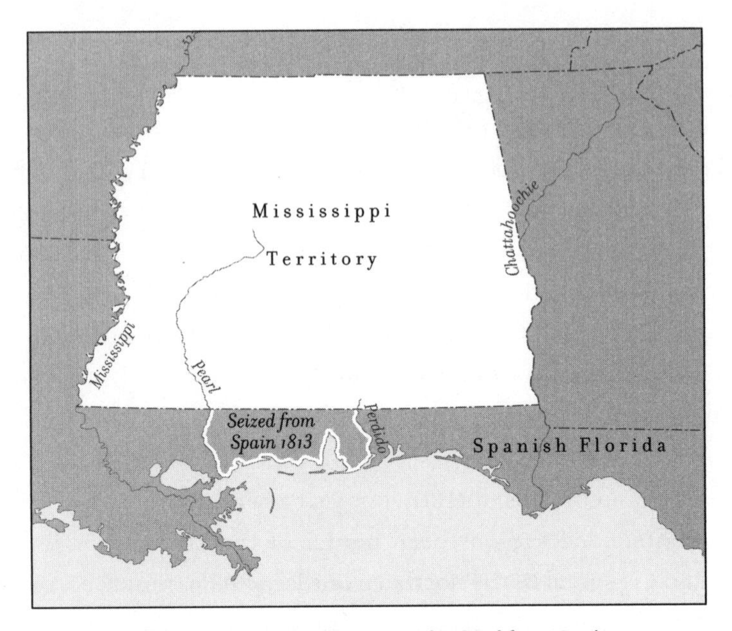

FIG. 102 Mississippi Territory—Land Seized from Spain

the creation of the state of Mississippi, the territory was divided almost perfectly in half along a vertical border. By vertically dividing the Mississippi Territory, both Mississippi and Alabama have a share of the rich bottomland to the south and their own access to the Gulf.

But why is their dividing line bent? And how come Alabama took a little snip from Mississippi's northeast corner?

The little snip up in the corner is the result of the Tennessee River. Had the vertical boundary simply continued, Mississippi's northeast corner would have been a tiny triangle isolated from the rest of the state by a river. Rather than try to govern an area with such difficult access, Congress made this segment of the Tennessee River the border, giving Alabama this "snip" to which it has direct access. (See Figure 19, in ALABAMA.)

The bend in the line dividing Alabama from Mississippi reflects the value of the rich bottomland soil and the rivers flowing directly to the Gulf of Mexico in the southern region of the territory. Because it was so highly prized, Congress divided this section equally along a line due south from the northwest corner of what was then Washington County.

But to achieve an equal division of the entire territory, Congress then angled the remaining boundary from the northwest corner of Washington County northward to the point where Bear Creek meets the Tennessee River, with a small segment of the Tennessee River completing the border. (See Figure 18, in ALABAMA). With this "bent" line, the difference in size between Mississippi and Alabama is less than one percent.

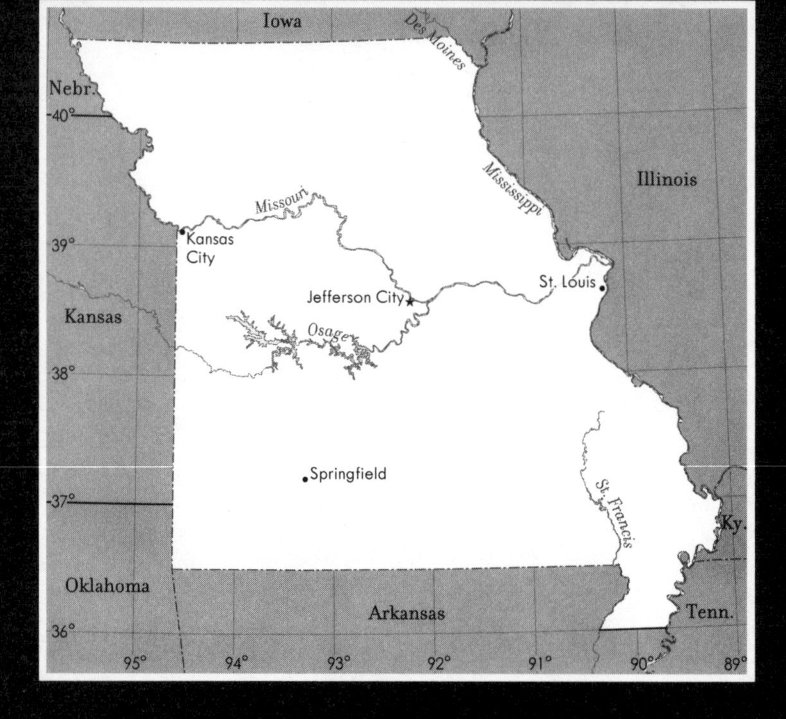

Why does Missouri's western border leave its straight-line path to follow a river, thereby messing up Kansas' perfect rectangle? Why does most of the southern border of Missouri line up with Kentucky's instead of Kansas's? And why is there that nib in Missouri's southeast corner?

Missouri's Eastern Border

The land that is now known as Missouri came into the possession of the United States as part of the Louisiana Purchase (1803). When President

Thomas Jefferson bought the land from France, few of its boundaries were precisely known. One of its borders that was clearly recognized was its eastern border, the Mississippi River, which until then had been the western border of the United States. That is why Missouri's eastern border is the Mississippi River.

Missouri's Southern Border

In 1817, residents in the regions of the Missouri and Mississippi rivers first petitioned Congress to create a state of Missouri. Over the following year, two more petitions were forwarded to Congress. Of the three petitions, two were from citizens' groups and the other from the territorial legislature. They differed as to their proposed western and northern borders, but all three proposed a southern border located at 36°30'. This boundary was viewed as simply an extension of the Virginia–North Carolina/Kentucky-Tennessee border. But it was about to become something more. (See Figure 10, in DON'T SKIP THIS.)

Back in 1787, Congress had prohibited the introduction of slavery into the newly designated Northwest Territory or any state created from that land. Many Americans believed the same prohibition ought to be applied as well to the Louisiana Purchase (1803), of which Missouri was a part. But the southern states could see that if slavery was prohibited in the states created from the Louisiana Purchase, slave states would soon be so outnumbered in Congress that slavery might be made unconstitutional. Missouri, as it turned out, was the first state to be created from the Louisiana Purchase. (Louisiana itself had been "grandfathered" into the Union along with its preexisting slaveholding status.) With Missouri, therefore, this question was put to the test.

Ultimately an agreement was reached known as the Missouri Compromise (1820). Under its terms, slavery would be prohibited in those parts of the Louisiana Purchase north of 36°30', with the exception of Missouri. (See Figure 25 in ARKANSAS.) This boundary line created a nearly equal division of the country between the regions where slavery would be permitted and the regions where it would be prohibited—given that the Rocky Mountains were poorly suited for settlement and given that several slave states already existed above 36°30'.

Not all of Missouri's southern border, however, is located along 36°30'. The eastern end of its southern border, known as the "boot heel," digs into what one would expect to have been Arkansas. The boot heel is living witness to what one man can achieve, given the right opportunities. In 1811, the eastern end of this area experienced a significant and frightening earthquake. Many of its residents moved away. But not John Hardeman Walker, a bold young man of seventeen, who recognized an opportunity when he saw one. Walker remained and acquired a great deal of the vacated land. In time he was known as the "czar of the valley"—and a very fertile valley at that, irrigated on the west by the St. Francis River and on the east by the Mississippi River (the eastern and western borders of the boot heel). When Walker discovered that the southern boundary being considered for the new state of Missouri would put his land in Arkansas, he took action.

Two things Walker understood well were land and power. Arkansas lacked the natural resources of Missouri, where two major rivers, the Missouri and Mississippi, connected at St. Louis. To be powerful, Walker knew, one needed to be connected. And Walker was. Even now we can see evidence of his power in the boot heel of Missouri that Congress allowed as an exception to Missouri's southern border.

Missouri's Western Border

The petitions for Missouri statehood that came from the citizens' groups proposed a western border along the Osage boundary line, which ran from Fort Clark on the Missouri River due south to the Arkansas River. The territorial legislature, however, proposed a border farther west that included land recently granted to the Sac and Fox Indians in return for, among other things, Iowa. (Figure 103)

Congress denied the Missouri legislature the additional western land

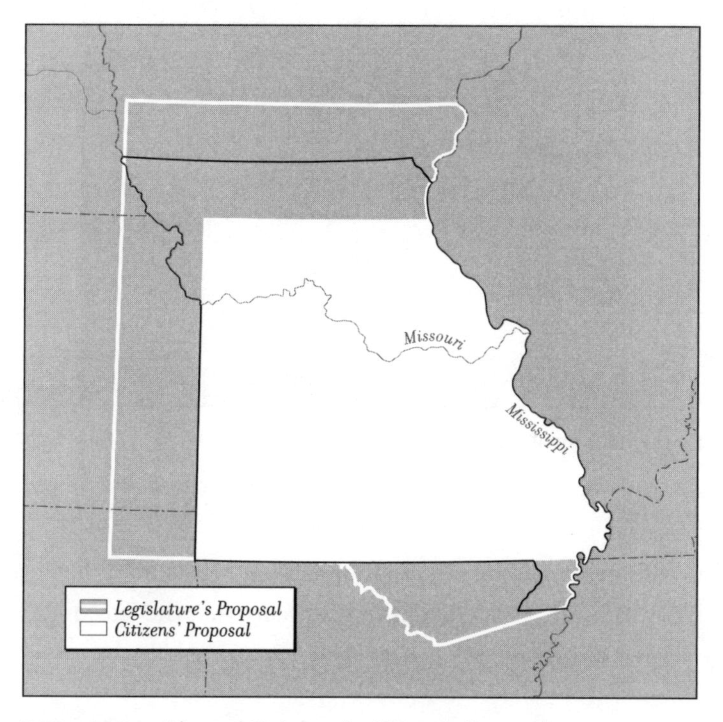

FIG. 103 Missouri Statehood—Alternate Proposals

it sought, creating the state with its western boundary along the Osage treaty line. After becoming a state, however, Missouri again sought to expand its western border. This time around, Missouri limited its bid to gaining more frontage along the Missouri River. (Figure 104)

This idea created two problems. Missouri was already one of the largest states in the Union. Adding additional land would further undermine the principle that all states should be created equal. Second, the land Missouri wished to annex belonged to Indians. But Congress was of many minds about Missouri. Though concerned about its size, southern congressmen recognized that the bigger Missouri was, the less room there would be for the creation of more (free) states north of 36°30'. From the point of view of northern congressmen, should the Missouri Compromise ever become void (and in time it did), the less room there would be for more slave states.

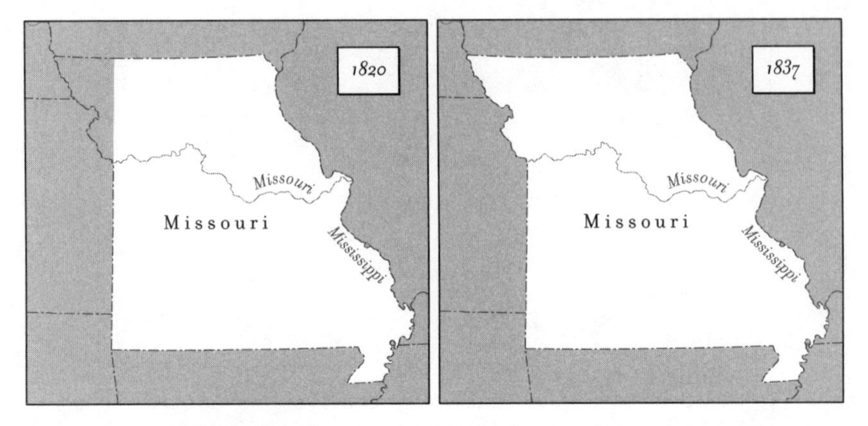

FIG. 104 Additional Land Annexed to Missouri

While Congress had mixed emotions about Missouri, it had no emotions about Indians, as is evident in the treaty that the Sac and Fox ultimately signed in 1836:

> Now, we the chiefs, braves, and principal men of the Sac and Fox tribes of Indians, fully understanding the subject, and well satisfied from the local position of the lands in question, that they can never be made available for Indian purposes, and that an attempt to place an Indian population on them must inevitably lead to collisions with the citizens of the United States; and further believing that the extension of the State line in the direction indicated, would have a happy effect, by presenting a natural boundary between the whites and Indians; and, willing moreover, to give the United States a renewed evidence of our attachment and friendship, do hereby, for ourselves, and on behalf of our respective tribes forever cede . . . the lands lying between the State of Missouri and the Missouri river.

President Martin Van Buren promptly proclaimed the land part of Missouri, giving the state the western border it has today.

Missouri's Northern Border

While Missouri's southern border has become, historically, its best-known border, its northern border was originally the source of the most passionate debate. The northern border of Missouri was marked off by surveyor John C. Sullivan in 1816. Sullivan's line ran due north from the juncture of the Kansas and Missouri rivers for a distance of 100 miles, forming a western border. It then turned east, forming the northern border, ending at the Des Moines River.

Five years later, when Congress created the state of Missouri, it specified a straight-line northern border corresponding to that surveyed by Sullivan. Except Sullivan's line wasn't straight. It veered northward at its eastern end. When the Iowa Territory came into being in 1838, it questioned this discrepancy. Missouri maintained it was an adjustment to accommodate the line's western end, 100 miles north of the Kansas and Missouri river juncture, and its eastern end, "the rapids of the river Des Moines" (a phrase used in Indian treaties that were written *after* Sullivan had surveyed his line).

Iowa was skeptical. For one thing, there were no rapids in the Des Moines River. (See Figure 70, in IOWA.) And who, Iowa wondered, had authorized Sullivan to survey this line? The answer surfaced in 1838, when Iowa asked the federal government to investigate. (For the details, go to IOWA.)

Ultimately, the U.S. Supreme Court ruled on the dispute. The Court recognized that the line's origins and execution were less than the best. But in the interim, numerous Indian treaties had been concluded that used this border, along with portions of land that had been deeded in the region by the state of Missouri. The Court therefore ruled in favor of Missouri, and Sullivan's line remains, to this day, the northern border of Missouri.

Each of Missouri's borders is evidence of events in which important principles were played out. Missouri's southern border preserves key events regarding slavery. Also, this border's "boot heel" remains as a legacy to the impact of individual power. Missouri's northern border preserves artifacts of raw power, as it gave Missouri more than its share of land and of frontage on the Mississippi River compared to Iowa on its north and Arkansas on its south. That raw power originates in the fact that Missouri possesses St. Louis, the juncture of two of 19th-century America's vital arteries, the Mississippi and Missouri rivers.

As was true of a handful of states, Missouri was "more equal" than most. And the northern end of its western border, torn as it was from the Indians, preserves the fact that inequality generates greater inequality.

MONTANA

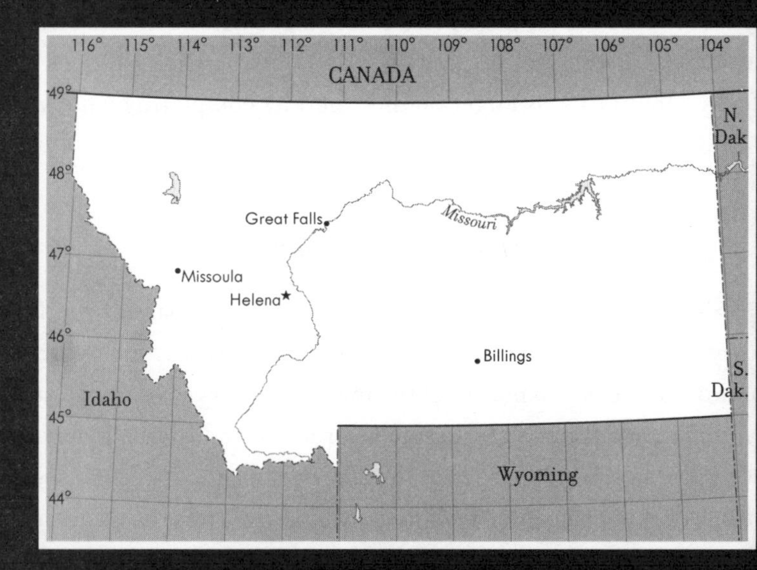

Why is there a sudden right angle downward in Montana's southern border? Wouldn't it have made more sense for that border to be entirely its straight east-west line? Why is Montana's eastern edge located where it is? And couldn't Montana have been divided down the middle and still have had enough area for two states?

Montana's Northern Border

The first of Montana's borders to surface—long before there was any idea of Montana—was its northern border. But it didn't surface over Montana. In the Convention of 1818, the United States and England specified that the northern extent of the Louisiana Purchase (1803) would be at 49°. (To learn why the 49th parallel was chosen, see DON'T SKIP THIS.) Nearly thirty years later, the United States and England divided the Oregon Country (see Figure 142, in OREGON) by extending this line from the Rocky Mountains to Puget Sound, thereby completing what would become Montana's northern border.

Montana's Eastern Border

Montana's eastern border first surfaced when it was part of the Idaho Territory, despite the fact that Idaho, as we know it today, sits to the west of Montana! The eastern border of the Idaho Territory was located at 104° W longitude. By locating the eastern border of the Idaho Territory at 104°, Congress reduced the size of the Dakota Territory to seven degrees of width. Congress also employed the 104th meridian as the eastern border of Wyoming, yet another state that eventually emerged with seven degrees of width. (See Figure 61, in IDAHO.) Three other western states, Washington, Oregon, and Colorado, also have seven degrees of width. The location of Montana's eastern border, then, has to do with the width of six states—none of which is Montana. (For more on discontinuous multistate borders, see DON'T SKIP THIS.)

Montana's Southern Border

While Montana's southern border did not appear until the creation of Montana in 1864, in effect the line had been drawn five years earlier when Colorado became a territory. This may seem strange, since Colorado does not border, and never has bordered, Montana or the Idaho Territory. When Colorado applied for territorial status, Congress adjusted its proposed northern and southern borders to their present locations. Seeking to create states that would be as equal as possible, Congress located these borders such that two additional states of identical height (four degrees) could fit between Colorado and Canada. Those two states turned out to be Wyoming and Montana. Thus Montana acquired a mathematically determined southern border. (See Figure 12, in DON'T SKIP THIS.)

Montana's Western Border

Montana's western border was created when the original Idaho Territory turned out to be impractical to govern. The Rocky Mountains in this region are so imposing that, at the time, it was virtually impossible to travel in winter between the eastern and western sides of the Idaho Territory. One year after creating the territory, Congress redefined Idaho's borders. In the process, Congress created Montana.

The separation between Idaho and Montana begins where the Continental Divide intersects the 111th meridian. It then follows the Continental

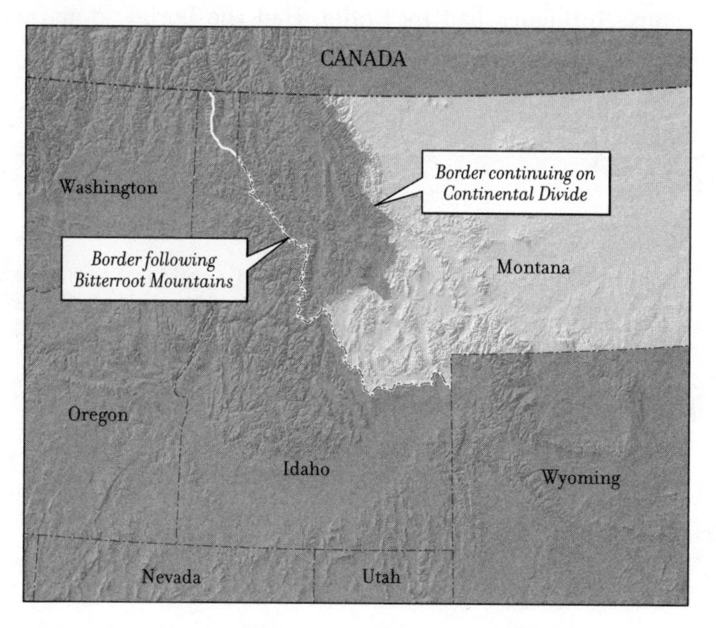

FIG. 105 Montana's Western Border—Two Proposals

Divide to the point where it intersects the Bitterroot Mountains. Here the crest of the Bitterroot Mountains becomes the boundary up to the Clark Fork River, where a straight line due north completes the border.

Why didn't Congress just stick with the Continental Divide? Especially since Idaho had proposed precisely this boundary. (Figure 105)

Among the delegates sent to Congress to urge approval for the creation of Montana was Judge Sidney Edgerton. Edgerton had felt himself snubbed by the governor of Idaho when, in the previous year, the governor assigned Edgerton to a judicial circuit east of the Rockies, which is to say, cut off from the action at the territorial capital. Edgerton turned out to be the wrong man to offend. A former congressman, he was personally acquainted with the chairman of the House Territorial Committee and with President Abraham Lincoln. Through Edgerton's efforts, Montana's western border pushed Idaho back from the Continental Divide to the Bitterroot Mountains.

Why, then, doesn't Montana's western border stick with the Bitterroot Mountains? Why does it switch to a straight line north? (See Figure 64, in IDAHO.)

Edgerton's influence had its limits. Had the border remained the Bitterroot Mountains, Idaho would have been deprived of the fertile

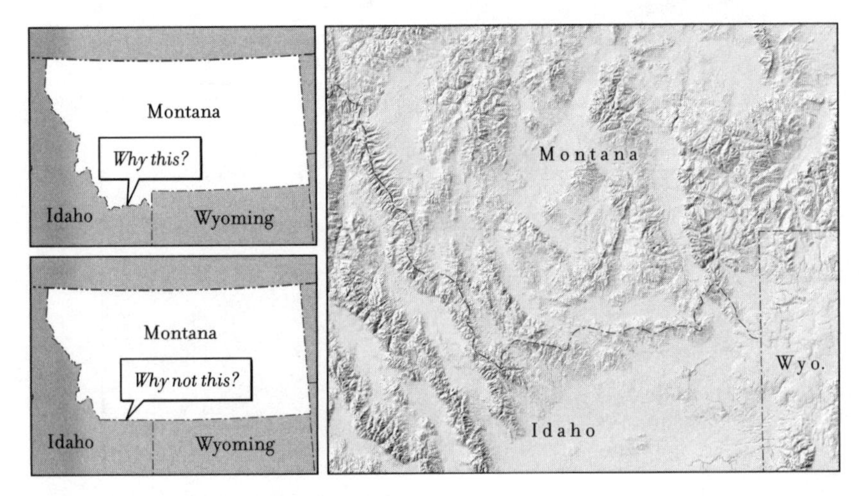

FIG. 106 Montana's Southwest Corner

Kootenai Valley and those valleys connected to it. While Montana did not lack agricultural land, Idaho needed every acre it could get.

But why does Montana's western border, down at its southern end, execute a U-turn? Wouldn't it have been simpler simply to end it upon reaching 45°, the latitude of its southern border? (Figure 106)

The reason for the western border's roundabout route is the result of the topography of southwest Montana. By continuing along the Continental Divide, the border preserves the mountain valleys on both the Montana and Idaho sides. The seemingly more efficient line would have sliced across those valleys, giving jurisdiction to Idaho for a very limited amount land on the other side of some very high mountains.

Montana's southwest boundary reveals that what may appear to be efficient in theory is sometimes inefficient in practice.

NEBRASKA

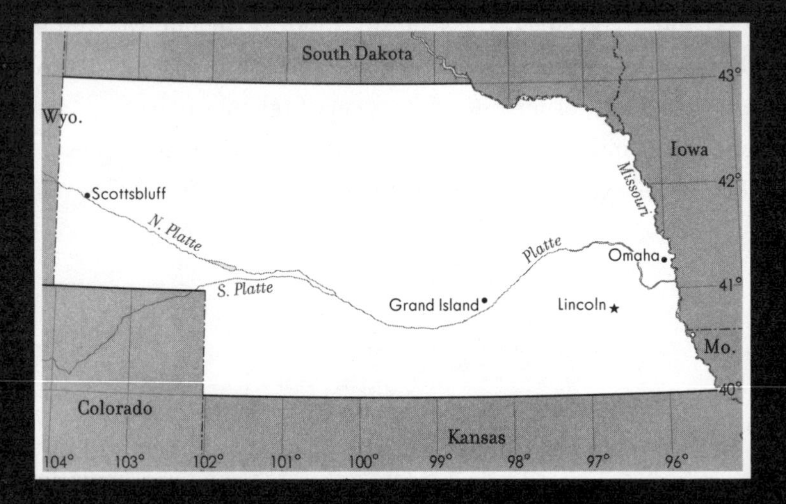

How come Nebraska lost its southwest corner to Colorado? And why are Nebraska's northern, southern, and western straight-line borders located where they are?

Nebraska looks like a nice enough state, tucked away in the middle of America. Who would suspect that in its younger and more territorially flamboyant days, Nebraska contributed to the borders of Colorado, Wyoming, Montana, North Dakota, and South Dakota?

Nebraska's Southern and Northern Borders

In 1854, the Kansas-Nebraska Act created the Nebraska Territory out of the Louisiana Purchase (1803) lands that were north of the 40th parallel. While this event may sound routine, the Kansas-Nebraska Act was, in fact, a major national upheaval. (For details on its violent events, go to KANSAS.) In the course of the conflict, Congress adjusted the proposed southern border of Kansas from 36°30' to 37°. By doing so, it not only resulted in Kansas having three degrees of height, but also created the opportunity for three more states with three degrees of height to fit between Kansas and Canada. Those states became Nebraska, South Dakota, and North Dakota. (See Figure 11, in DON'T SKIP THIS.) This, then, is not only when the future state of Nebraska acquired its southern border at 40° N latitude, but also when its future northern border was determined.

Nebraska's Eastern Border

Nebraska also acquired its present-day eastern border right from the outset, since it had already been established by its neighbors to the east, Missouri and Iowa. The Missouri River formed the western border of both these states in the region east of Nebraska.

Initially, the Nebraska Territory was huge. It extended north all the way to Canada and west all the way to the Continental Divide in the Rocky Mountains. (Figure 107)

The citizens of the Nebraska Territory never envisioned these boundaries as being those of a future state. Still, some very valuable assets were released in the years ahead to territories formed from Nebraska. For example, gold mines in the southwest corner of the territory were released without protest to create Colorado. Why?

Nebraskans knew that gold mines were not a box of chocolates. As in other territories where gold had been discovered (notably Kansas and Washington), the culture of those in the mining community was quite different from that of the earlier, agricultural settlers. And the vast

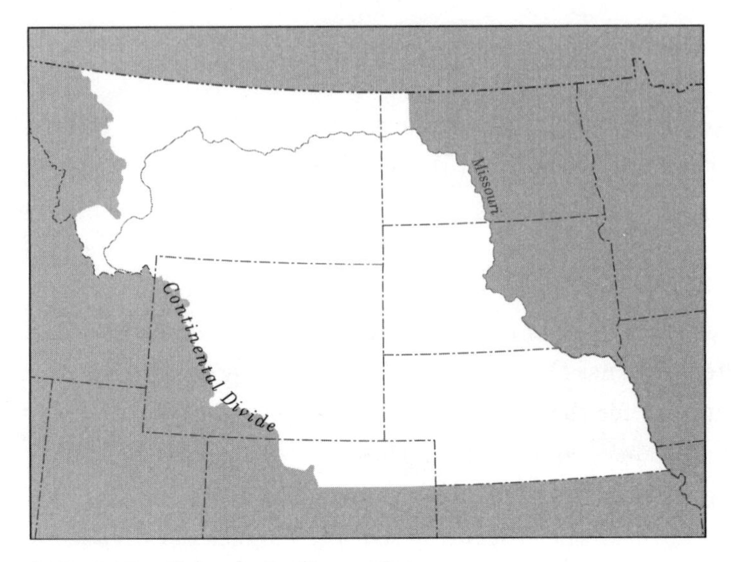

FIG. 107 Nebraska Territory—1854

numbers of mining people who flooded into these regions threatened the established power structure. It was thought to be better, perhaps, that they have their own territory.

Moreover, a different economic asset captured the imagination of Nebraskans. In 1861, Nebraska's territorial governor, Alvin Saunders, revealed that particularly prized asset in his opening address to the legislature:

> A mere glance at the map of the country will convince every intelligent mind that the great Platte Valley, which passes through the heart and runs nearly the entire length of Nebraska, is to furnish the route for the great central railroad, which is to connect the Atlantic and Pacific States and Territories.

With its southwest corner of mountains (which are not good for railroads) and miners (which are not good for farmers), Nebraska released this corner to Colorado with a smile.

The Nebraska Territory lost its largest chunk of land in the creation of the state of Minnesota. But Minnesota was never part of the Nebraska Territory. How, then, did it cost Nebraska land?

To become a state, the Territory of Minnesota released its land west of the Red River of the North. Congress then reorganized the territorial land in the center of the country. The region left over from Minnesota was combined with the bulk of what had been the Nebraska Territory to form the Dakota Territory. (Figure 108)

At the same time that Congress created the Dakota Territory, it was also seeking to further contain the Indians in order to enable the expansion that was following the railroads. One of the tribes Congress sought to contain was the Ponca Indians. The Ponca were not a militant people, but they were nomadic, which conflicted with the culture of settlers. The government concluded a treaty with the Ponca, confining them to an area whose borders influenced the new northern border of Nebraska. When the Dakota Territory was formed, Nebraska's northeastern border became the Missouri River. But rather than follow it to the 43rd parallel, the border diverted south along the Niobrara and Keya Paha rivers and joined the 43rd parallel farther west. The land between these rivers and the Missouri was reserved for the Ponca. (Figure 109)

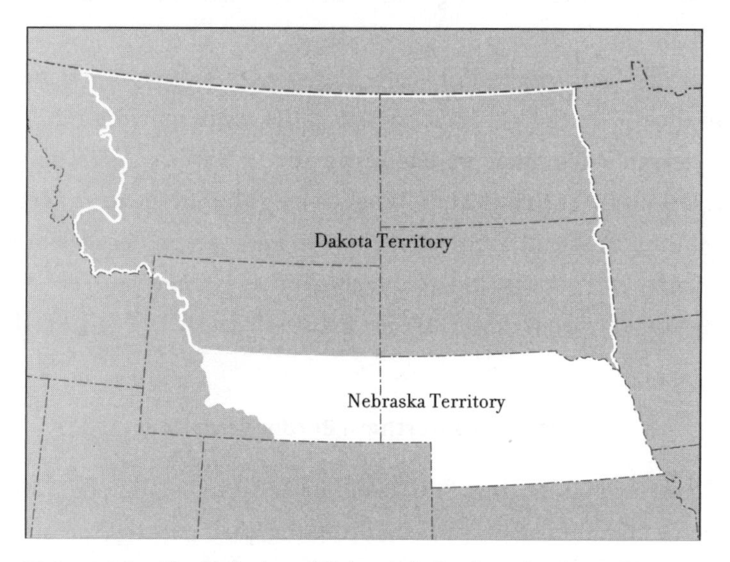

FIG. 108 The Surfacing of Nebraska's Northern Border—1862

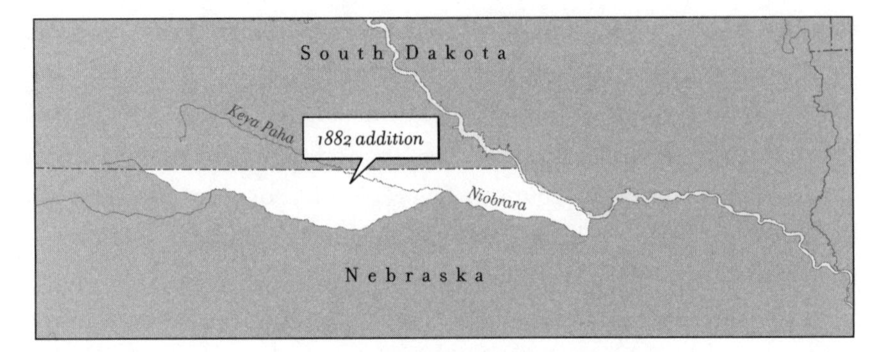

FIG. 109 Nebraska's Northern Border—Adjustment After Statehood

Nebraska's Western Border

In 1863, the territory of Washington released its mining lands for the same reasons that Kansas and Nebraska had released theirs. Congress combined these mining areas of the Washington Territory with the open plains eastward up to 104° W longitude, thereby creating the Idaho Territory. Though the Idaho Territory was redefined the following year, its significance for Nebraska resides in the fact that the Idaho Territory's eastern border of 104° W longitude remains to this day the western border of Nebraska.

Why 104°? By lining up the western borders of Nebraska and its neighbors to the north, the two Dakotas, along the 104th meridian, North and South Dakota each ended up spanning almost exactly seven degrees of width. Likewise, by lining up Nebraska's neighbor to the west, Wyoming, along the 104th meridian, it, too, came to span seven degrees of width. These three states ultimately joined with three other western states in sharing equal degrees of width. (See Figure 13, in DON'T SKIP THIS.)

Nebraska's Northern Border Revisited

In 1882, fifteen years after Nebraska had achieved statehood, Congress adjusted its northern border by making the 43rd parallel the boundary from the western end of the state to the Missouri River. (Figure 109) This adjustment remains today as evidence of a political chess game—in this case, a *deadly* political chess game—that the government was engaged in with Native Americans. The dominant tribe in the Dakotas was the Sioux. Their military prowess was such that in 1876 they wiped out General George A. Custer's 7th Cavalry. Not long after, the United States signed a treaty with the Sioux, granting them land that included the area previously set aside for the Ponca. In short order, the Sioux attacked the Poncas in a series of deadly raids. Some ten years later, the rejuvenated 7th Calvary massacred the Sioux at Wounded Knee, South Dakota. The revised northern border of Nebraska preserves one piece of this unhappy history.

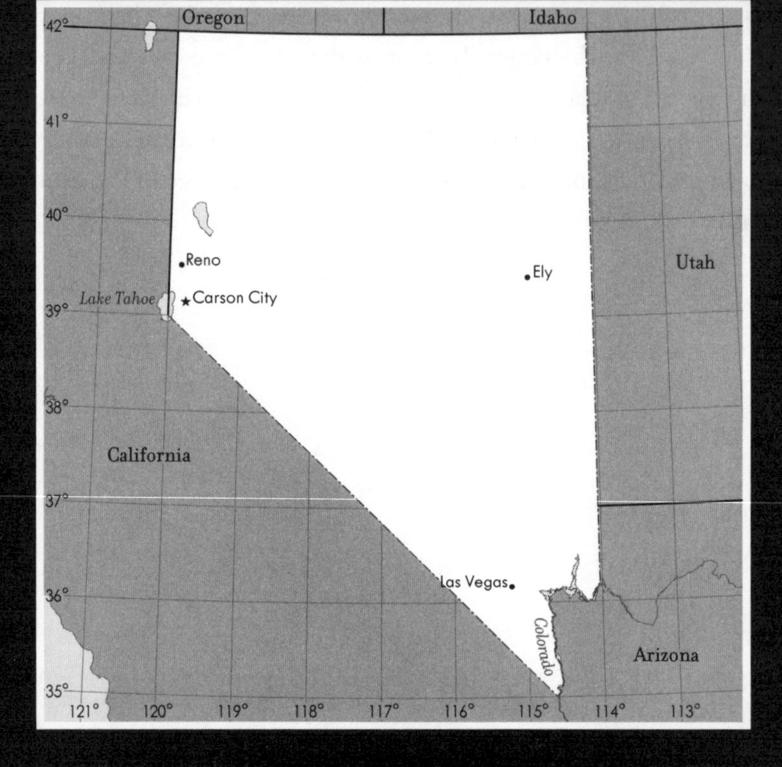

Is there any question as to why Nevada has the borders that it has? Aren't they obviously just straight lines that follow the lines established by its neighboring states? (Guess again.)

Nevada's Northern Border

Nevada acquired its northern border even before the United States acquired the land that would become Nevada. In 1790, England and Spain negotiated the Nootka Convention, which divided their territorial claims west of the Rocky Mountains along the 42nd parallel. (To find out why the 42nd parallel was chosen, see DON'T SKIP THIS.) Thus, when the United States acquired the land that includes present-day Nevada as part of its conquests in the Mexican War (1846–48), the 42nd parallel was already in place as its northern border.

Nevada's Western Border

When California became a state in 1850, just over a year after coming into American possession from the Mexican War, its controversial eastern border was established as two straight lines encompassing all of the gold-filled Sierra Nevada. (For more on this, go to CALIFORNIA.) The United States divided the remainder of its Mexican War acquisitions into the territories of Utah and New Mexico, the western borders of which were the California border.

During the first decade of its American existence, few people were interested in living in the western regions of the Utah Territory, the part that would become Nevada. It was mostly desert, called the Great Basin. But in 1859, a vast network of silver deposits, known as the Comstock Lode, was discovered in the western mountains of the Utah Territory. Soon after, more discoveries were made, not only of silver but also of gold. At the same time, the nation's conflict over slavery was erupting into war. As Americans began to pour money and lives into the Civil War, here in these barren hills in the western reaches of the Utah Territory was an untapped treasure of silver and gold.

Washington worried about the Utah Territory. Mormons constituted its main population, and their relations with the government had always been strained over the issue of polygamy—and, in a larger sense, over the issue of authority. To preserve possession of these silver deposits, Congress created the territory of Nevada in 1861. California, which had been the western border of the Utah Territory, was now the western border of Nevada.

Nevada's Eastern Border

The borders specified in the legislation creating the Nevada Territory were designed to separate the silver and gold mines from the Utah Territory. But discoveries of gold were continuing to be made. Within a year, it was evident that considerable gold remained just beyond Nevada's eastern boundary, the 116th meridian. Nevada's territorial delegation to Congress proposed that its eastern border be relocated one degree farther east. Congress complied. (Figure 110)

When Nevada was granted statehood in 1864, its eastern border was relocated yet another degree farther east to the 114th meridian. It had turned out there were more gold deposits east of Nevada's revised boundary, but this time there had been something else. Rivers.

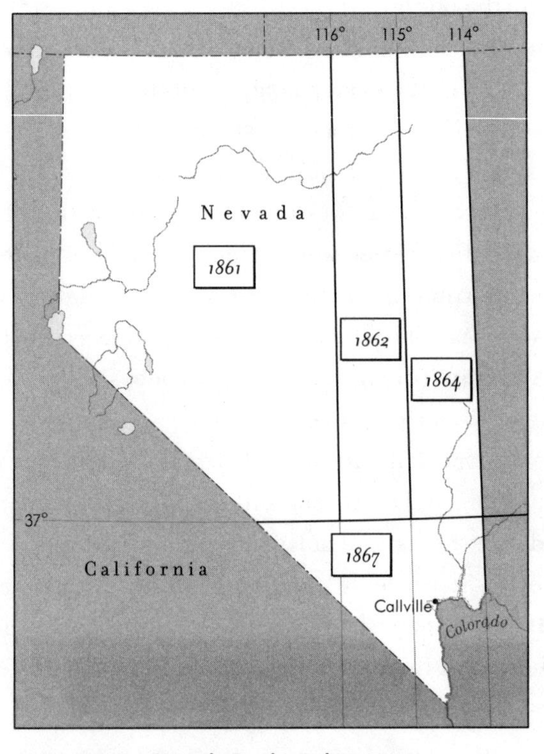

FIG. 110 Nevada Border Enlargements—1862–1867

A significant shift is evidenced by this boundary change. Initially, the determining factor in locating Nevada's borders had been what the government didn't want to be Utah. But in this instance, the border was relocated because of what Nevada did want to be Nevada: water. A land where no one had wanted to be was now a place where people did want to be.

Nevada's Southern Tip

Even after becoming a state, Nevada's borders were not through growing. In 1866, steamboat navigation was opened on the Colorado River from the Gulf of California as far as the town of Callville, in the Arizona Territory. Today Callville is located at the bottom of Lake Mead, created in 1935 by the Hoover Dam. Nevada, having no navigable river, could dearly use Callville for transportation to the sea. Again Congress complied. In 1867, it ceded more than 18,000 square miles of what had been the Arizona Territory to the state of Nevada. (Figure 110)

The Arizona Territory vigorously opposed what it viewed as theft. In its formal protest to Congress, its legislature stated why this region was so important to its citizens:

> It is the watershed of the Colorado River into which all the principal streams of Arizona empty, and which has been justly styled the Mississippi of the Pacific. By this great river the Territory receives the most of its supplies, and lately it has become the channel of a large part of the trade of San Francisco with Utah and Montana.

All true. Also true was that Arizona had aligned with the Confederacy during the recent Civil War. Not only did Arizona fail to persuade Congress to revoke this transfer of its land, but, later that same year, Congress added another chunk of Arizona to Nevada, this time the entirety of Arizona's westernmost end, giving Nevada its southern tip and the borders we know today. (See Figure 23, in ARIZONA.)

Initially, Nevada had come into existence as a result of the government's distrust of the Mormons. Non-Mormons feared that the Mormons undermined the nation by a lack of allegiance to its authority. That this nation was founded on the very notion of limiting governmental authority increases the complexity of this issue. But none of that complexity resides in the borders of Nevada. Rather, Nevada's borders remain today as evidence of authority pure and simple.

NEW HAMPSHIRE

Why does New Hampshire have a straight-line eastern border, and

New Hampshire's Eastern Border

The place we know as New Hampshire was previously part of the Plymouth Colony, the forerunner of what was to become Massachusetts. In 1622, Sir Fernando Gorges, a principal stockholder in the Plymouth Company, formed a partnership with Captain John Mason under which the two obtained proprietorship for all the land between the Merrimack and Kennebec rivers. (See Figure 82, in MAINE.) Mason and Gorges dissolved their partnership in 1629, with Mason taking as his share those lands west of the Piscataqua River and calling this land New Hampshire. The Piscataqua River remains, to this day, the southernmost segment of New Hampshire's eastern border.

New Hampshire's Southern Border

As it turned out, New Hampshire was a bad business venture. Mason's heirs abandoned it. But that did not stop settlers from moving into the land, and in 1639 many of them joined in a compact for governance that stated they would live by the laws of Massachusetts. Other settlers believed New Hampshire should become its own colony. They, in turn, signed a compact in which they pledged

> to submit to his royal Majesty's laws, together with all such laws as shall be concluded by a major part of the freemen of our society, in case they be not repugnant to the laws of England, and administered in behalf of his Majesty. And this we have mutually promised, and engaged to do, and so to continue till his excellent Majestic shall give other orders concerning us.

If this sounds like flattery, that's because it is. But they had a reason for flattering the king: Massachusetts. This powerful colony was composed primarily of Puritans, who were increasingly at odds with the king. (Three years after New Hampshire's plea for self-rule, England's Puritans participated in the overthrow of the monarchy.) In New Hampshire, many colonists reasoned that the king might like the idea of weakening the Puritans' power in America by hacking off a hunk of Massachusetts.

And in fact the king did like it. Though not the king they wrote to. He had been beheaded. But in 1680, his son, restored to the throne by Parliament, decreed an independent government for New Hampshire. But in the decree, Charles II stopped short of stipulating its borders.

Undeterred, New Hampshire claimed its own borders. New Hampshire sought a southern border that began 3 miles north of the mouth of the Merrimack River, the boundary specified in the charter of the Massachusetts Bay Colony as its northeast corner. From this point, New Hampshire claimed its border with Massachusetts was a line due west.

Massachusetts disagreed. Based on its royal charter, it maintained that its border was a line 3 miles north and east of the Merrimack River. (Figure 111)

Not until 1741 was the dispute resolved, when King George II decreed that the boundary followed, for the most part, the one claimed by New Hampshire. It paralleled the Merrimack River at a distance of 3 miles to the north (as the Massachusetts charter specified), but at the town of Lowell, where the Merrimack River turns north, the line drawn by

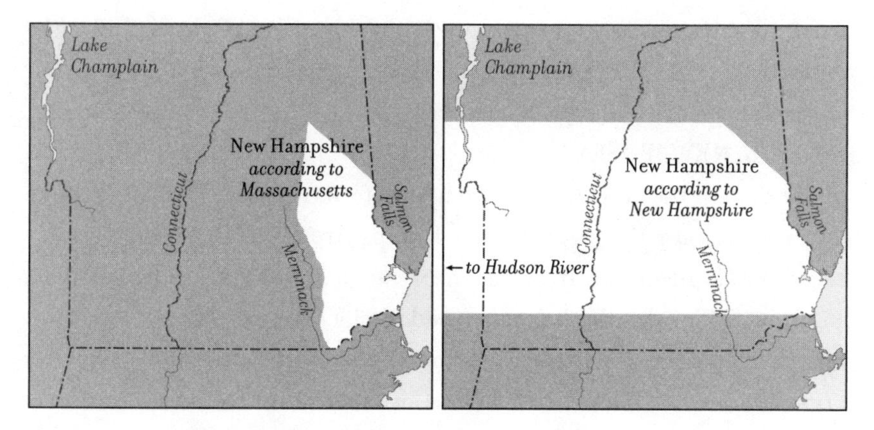

FIG. 111 New Hampshire/Massachusetts Border Dispute—1740

George II continued due west. This line (though subsequently surveyed inaccurately) has remained the southern border of New Hampshire to this day. (See Figure 92 in MASSACHUSETTS.)

New Hampshire's Eastern Border (Revisited)

Throughout this time, New Hampshire's settlements had been expanding. Recognizing this, George II dispatched commissioners to New Hampshire to survey an extension of the colony's eastern border. In 1737, the commissioners' report declared New Hampshire's eastern border would now extend from the headwaters of the Salmon Falls River "north two degrees west" until reaching a distance of 120 miles from the mouth of the Piscataqua River. The designation of two degrees west of north was used to finesse the fact that, on a flat map, a line that is due north would appear increasingly curved the farther it was from the map's center meridian. While it is an imperfect approach, it seems to have functioned well enough, for this line has remained in place ever since. (See Figure 84, in MAINE.)

New Hampshire's Western and Northern Borders

New Hampshire assumed its western border was that claimed by its predecessor, Massachusetts: the Hudson River and, to its north, Lake Champlain. New Hampshire's western neighbor, New York, also believed its border was that of its predecessor—the Dutch—who had claimed their eastern boundary to be the Connecticut River. (Figure 112. See also Figure 114, in NEW JERSEY.)

Once again the conflict was brought before the king. This time, however, the crown ruled against New Hampshire. In 1763, George III declared the Connecticut River was the western border of New Hampshire. (Figure 113) This border has remained in place ever since.

Why would the king give this land to New York, a colony already far larger than New Hampshire? The king's decision reveals a significant difference in the way borders were determined before and after the

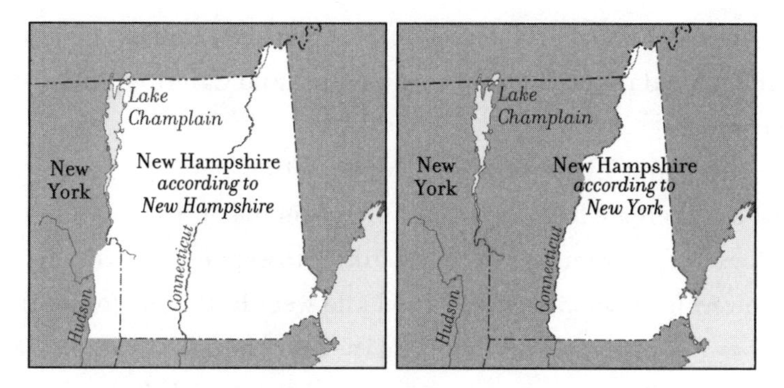

FIG. 112 New Hampshire/New York Border Dispute—1760

American Revolution. The king was not concerned with equality. Nevertheless, his decision also reveals a very perceptive political mind.

When George III denied New Hampshire the western border it sought in 1763, England had just come into possession of what is now called the province of Quebec. This gave England possession of the St. Lawrence River, a lifeline for its fur trade in the Canadian interior. It also gave England a lot of new subjects—French-speaking, Catholic subjects. The

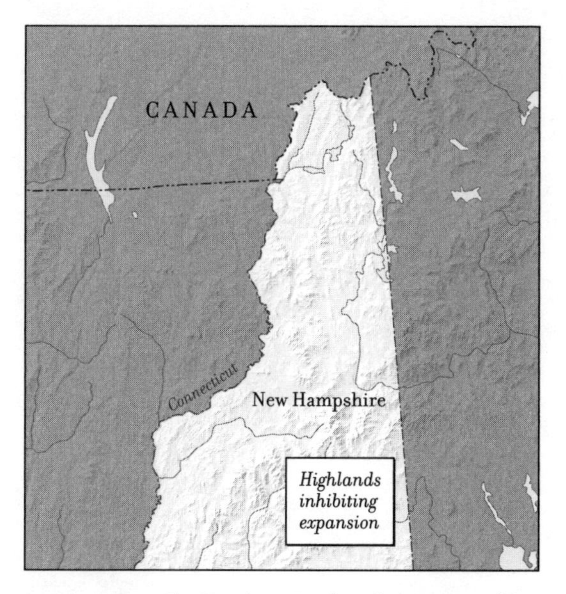

FIG. 113 The Northern Border of New Hampshire

Quebecois were not at all happy to find themselves under England's authority, particularly with those expansionist and Calvinist colonists just to their south.

The last thing England needed was conflict between these groups, since it could create havoc with commerce on the St. Lawrence. England knew New York's colonists were less likely to expand toward Canada, as their colony possessed plenty of land and wealth, thanks to the Hudson River. New Hampshire's colonists, on the other hand, had amply demonstrated that they were very much inclined to expand. By declaring the Connecticut River to be the western border of New Hampshire, England was seeking thwart any further expansion by New Hampshire, as this border fenced New Hampshire behind a river and behind sizable highlands between it and Quebec.

NEW JERSEY

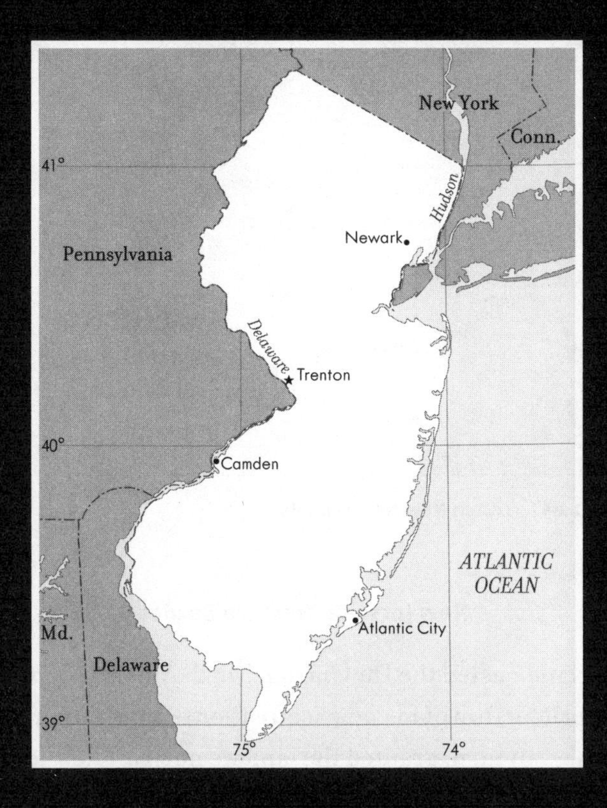

Who tilted New Jersey's northern border and why? Why not just straight across? And why, since it is tilted, does it start and end at the points where it does? Wouldn't it all make more sense if New Jers

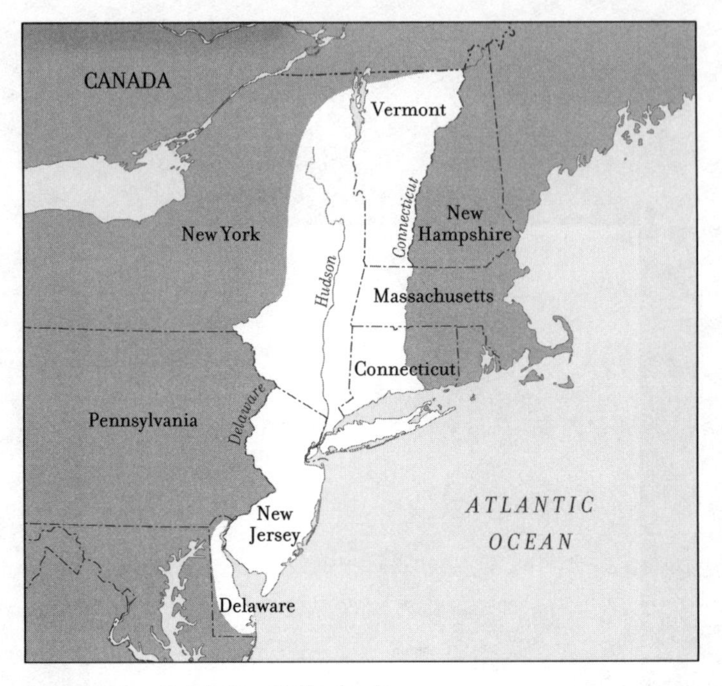

FIG. 114 **Dutch New Netherlands**

New Jersey's Western Border

In 1674, England ousted the Dutch from North America. King Charles II granted proprietorship of his new acquisition to his brother, the Duke of York. The duke, in turn, granted the land we now call New Jersey to Lord John Berkeley and Sir George Carteret. The duke defined the western border of New Jersey as being the shoreline along the Delaware River and Bay, which remains the western border of New Jersey to this day. With two exceptions. Twice, what would clearly seem to be New Jersey is actually Delaware! (See Figure 2, in INTRODUCTION. For more details, go to DELAWARE.)

New Jersey's Northern Border

The Duke of York defined the northern border of New Jersey as being the line between "the northernmost branch of the said bay or river of

Delaware, which is 41°40' latitude" and the point where the Hudson River crosses 41° N latitude. Thus, the Delaware River, which had been the western border of the Dutch, became and has remained the western border of New Jersey.

As for New Jersey's northern border, it would be some years before it was discovered that the northernmost branch of the Delaware is considerably north of 41°40'. (Figure 115)

That New Jersey is not simply the eastern end of Pennsylvania is not for lack of trying on the part of Pennsylvania. In 1681, William Penn and others from his colony purchased the western half of New Jersey from Sir George Carteret's widow. But confusion over authority led England to reorganize the region. In 1702, the crown united East and West Jersey into New Jersey and placed New Jersey under the rule of New York.

New Jersey finally became a colony in its own right in 1738, and one of its first actions was to survey its borders. This survey revealed the northern border contradictions in the royal grant to Carteret and Berkeley.

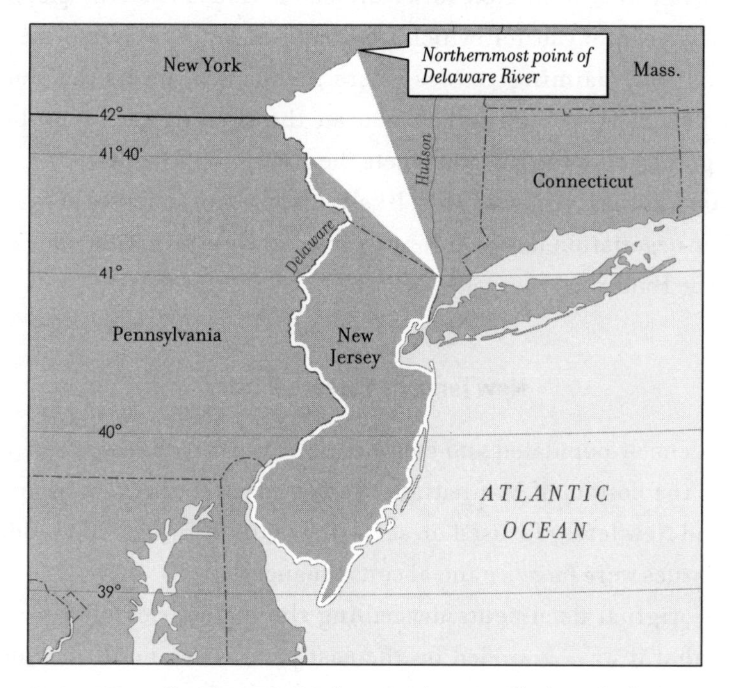

FIG. 115 New Jersey's Northern Border—Conflicting Stipulations

Graciously (or more likely, realistically), New Jersey limited its claims to the more conservative of the two locations mistakenly stated by the duke—that being the one locating its northwest corner at the point where the Delaware River crosses 41°40'. Even so, New Jersey now claimed possession of a narrow strip of land along the north side of the Delaware. Conflicts and violence ensued as authorities from the colonies of New Jersey and New York sought to collect taxes and record deeds for this strip of land. Only after the American Revolution did the two states agree to relocate New Jersey's northwest corner to its present location, which is the point where the Delaware makes a 90-degree turn to the northwest, thereby ceding to New York the narrow strip of land between the river and 41°40'.

Still, New York and New Jersey continued to spar over the other end of this border: the point where the Hudson River meets the 41st parallel. According to New Jersey, the Hudson met the 41st parallel at the town of Haverstraw, which in fact was so far north of 41° one wonders if New Jersey had very bad surveyors or very good negotiators. The latter appears to be the case, since when New York claimed the Hudson met the 41st parallel at the town of Closter, which was south of 41°, New Jersey met New York halfway, claiming 41° was where the Sparkill meets the Hudson. Ultimately, the two sides agreed to locate the eastern end of their border at the point where the Hudson meets the *actual* 41st parallel.

What was that all about? Most likely, it was about applying pressure in another negotiation between New Jersey and New York, this one farther down the Hudson.

New Jersey's Eastern Border

In the densely populated and commercially bustling harbor area in and around the bottom of Manhattan, the boundary conflicts between New York and New Jersey focused on seemingly minute issues. But hidden in those issues were huge financial consequences.

The original documents describing the boundaries of New Jersey stated that it was "bounded on the east part by the main sea and . . .

Hudson's river." Sounds simple enough. Unless someone asks if this description means that the Hudson River is part of New Jersey or comes up to the edge of New Jersey. New York took the position that the Hudson River met the edge of New Jersey, making the river itself within New York's jurisdiction. New Jersey took the position that its separation from New York's jurisdiction in 1738 entitled it to a border at the midpoint of the Hudson and the harbor into which it flowed. As such, New Jersey claimed its borders included any island to the west of that midline, such as Staten Island (which is, after all, much closer to New Jersey than to New York).

A long dispute ensued. Not until 1833 did the two states come to an agreement, and the only way they managed that was with some very creative borders. Under the agreement, the boundary of the land under the water was located at the center of the Hudson River and Upper New York Bay. A second boundary, on the surface of the water, gave New York jurisdiction over all land above the water up to the mainland. (Figure 116) Excepted from this, however, were "above water" surfaces that were

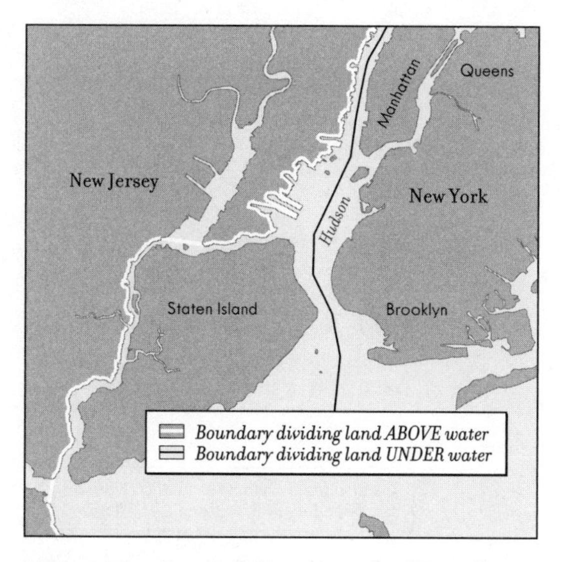

FIG. 116 New York/New Jersey Dual Boundary—1833

attached to the New Jersey mainland and anything attached to those surfaces. What were those lawyers talking about?

Docks. That is what New Jersey got out of this deal, along with possession of the land under the water on their half of the Hudson, for what that's worth. (And we'll soon see what that's worth.) But the immediate gain was that New Jersey got to have a harbor and that is the multimillion dollar reason the New York/New Jersey boundary acquired this peculiar dual course.

While New York won jurisdiction over the islands in Upper New York Bay (including, therefore, Staten Island), New Jersey appears to have had the last laugh. In 1892, so many immigrants were pouring into New York that it needed an enlarged facility for processing them. (At that time, immigration was handled by the states.) It selected Ellis Island in Upper New York Bay. The tiny island was enlarged with landfill to accommodate the structures required. In recent years, Ellis Island, no longer used for immigration, has become a very popular tourist site.

In 1993, New Jersey dusted off its 1833 agreement and headed to court, claiming that those areas of Ellis Island that had been created out

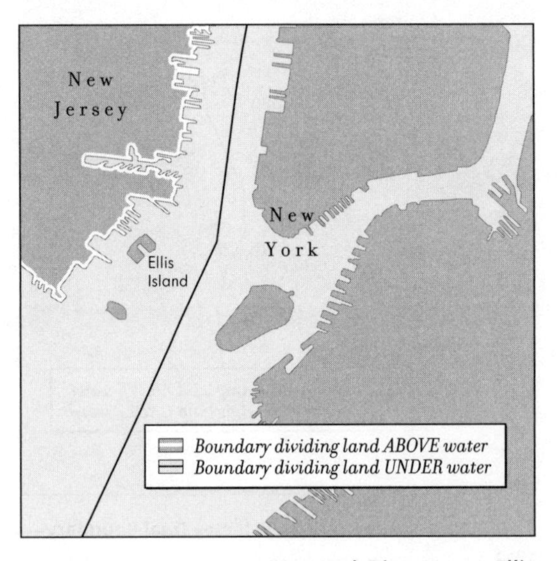

FIG. 117 New Jersey/New York Dispute over Ellis Island

of landfill actually belonged to them, since the agreement gave New Jersey the land under the water that was west of the mid-channel line. (Figure 117)

New Jersey argued that New York had trespassed on its underwater land and then, by filling it to a point above the water, added theft to trespassing! New York didn't view it that way. But the Supreme Court did, and in 1998 it ruled that those areas of Ellis Island that were below water in 1833 are today part of the State of New Jersey.

NEW MEXICO

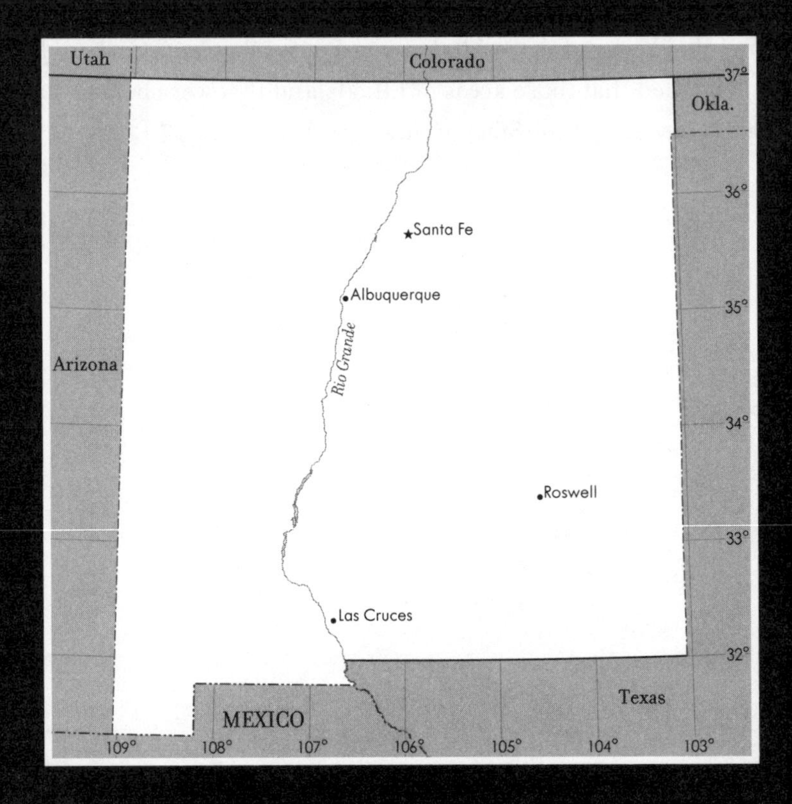

How come New Mexico has all those "steps" along its southern border? And why does this border almost, but not quite, line up with the border at this corner of Texas? Is there some reason why New Mexico's northern border lines up with Oklahoma's northern border instead of the border of Texas? New Mexico's western border with Arizona continues north as the border between Colorado and Utah—is there some reason why this four-state line intersection is located where it is?

FIG. 118 Mexican New Mexico

Most of what would become the state of New Mexico came into the possession of the United States as part of the land acquired from the Mexican War (1846–48). (See Figure 28, in CALIFORNIA.) Before the Mexican War, "New Mexico" was a province in Mexico called . . . Nuevo Mejico. (Figure 118)

New Mexico's Southern Border (western end)

The only part of Nuevo Mejico's borders that remains today is an echo of its southern border. When the United States created the territory of New Mexico in 1850, it was in conjunction with the Compromise of 1850, part of which entailed the United States buying land from Texas and giving it to the new territory. The southern border of this purchase was on the same line as what was then the southern border of the New Mexico Territory, and prior to that, of Nuevo Mejico. Three years later, the Gadsden Purchase wiped out the original southern border of Nuevo Mejico, though its echo remains in the southern border of that portion of New Mexico

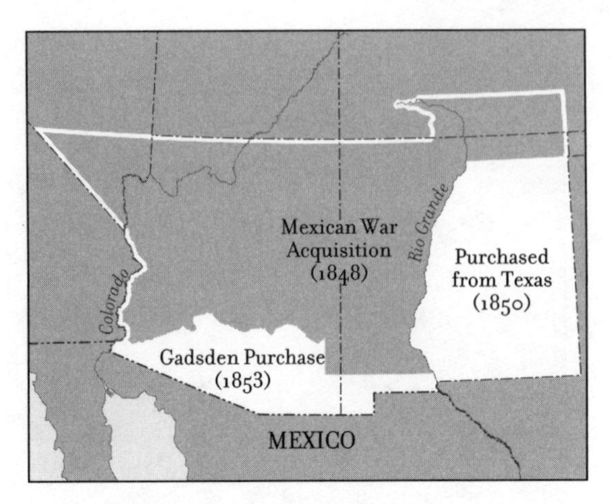

FIG. 119 New Mexico Territory

purchased from Texas. The southern side of the Gadsden Purchase constitutes the rest of New Mexico's current southern border. (Figure 119)

The reason that part of New Mexico's southern border does not quite line up with the part that borders Texas is that the goal of the Gadsden Purchase had been to buy land from Mexico with adequate mountain passes for the United States to build a railroad. A critical passage for the railroad through the San Andreas Mountains (the city of El Paso is named for a pass on the route) resulted from the purchase of land that expanded a segment of New Mexico's southern border to just below its previous alignment with Texas. (For more details about the Gadsden Purchase, go to ARIZONA.)

New Mexico's Eastern and Western Borders

The acquisition of the land that is now known as New Mexico brought with it a challenge very much like the challenge President Thomas Jefferson faced following the Louisiana Purchase. Suddenly the nation had acquired a sizable population that spoke another language and possessed a different culture. Congress responded by continuing the approach to state borders recommended by Jefferson. It created a border designed to make New Mexico's Spanish-speaking population feel secure, represented, and equal.

The task of incorporating New Mexico's Spanish-speaking population did not go entirely smoothly. One element that helped, however, was that Texas desperately needed money. Its days as a republic had left it deeply in debt. This was why Texas had sold the United States a large section of its land in the Compromise of 1850. In the same legislation, the needs of the nation's new citizens in Santa Fe are evidenced by the location of New Mexico's new eastern border.

The eastern border of New Mexico is located three degrees of longitude from Santa Fe, putting the town about 160 miles from Texas, a comfortable distance from these powerful and ambitious Americans. But why 160 miles, why not 150 or 200? In this location, we can see Congress looking ahead to the time when the large New Mexico Territory would be divided (1863) for purposes of creating states of roughly equal size. The distance from Santa Fe to what will eventually be the New Mexico/Arizona border is almost precisely the same as the distance from Santa Fe to the Texas border. And the future state of Arizona is very nearly equal in width to that of the future state of New Mexico. (Figure 120) New Mexico

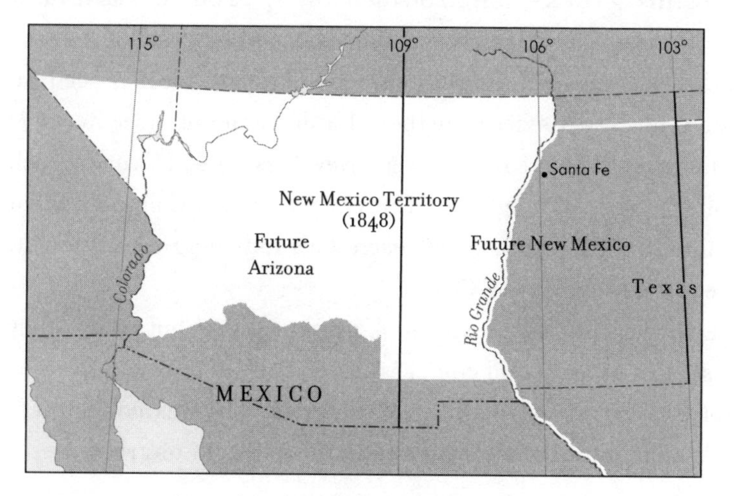

FIG. 120 The Logic Behind Eastern and Western Borders of New Mexico

and Arizona are today so nearly equal in size that, among all fifty states, they rank fifth and sixth, respectively.

New Mexico's Northern Border

Initially, the eastern end of the territory of New Mexico's northern border followed the 38th parallel to the San Juan Mountains, then followed the crest of these mountains south to the 37th parallel, at which point the 37th parallel continued as the rest of New Mexico's northern border. When the federal government added land to New Mexico from its 1850 purchase from Texas, it made the 37th parallel its northern boundary for the entire territory. That is the line that remains to this day.

But why the 37th parallel? One might expect this location had to do with the Missouri Compromise (1820), under which slavery was prohibited in newly created territories with borders north of 36°30'. But, in fact, Congress did not ban slavery in New Mexico. Under the Compromise of 1850, a passionately divided Congress decided to let the New Mexico Territory decide for itself whether or not to permit slavery. In doing so, Congress planted the seed for what was to become known as "popular sovereignty"—the country's next major effort to cope with slavery. (For more on this, go to KANSAS.) So then why 37° if 36°30' was invalidated? Why not line up New Mexico's northern border with that of Texas?

The answer is that Congress was also planting another seed in 1850. Using 37° for New Mexico's northern border, a tier of three Rocky Mountain states could be created between New Mexico and Canada, each having four degrees of height. And in years to come, Colorado, Wyoming, and Montana filled that space, each with four degrees of height. (See Figure 12, in DON'T SKIP THIS.)

The borders of New Mexico look pretty square, but in fact they preserve stories of fears and compassion, of shrewd political savvy, and of objective planning for the future. In a sense, New Mexico's borders contain a kind of mural of what goes on in the halls of Congress.

NEW YORK

Why does New York's eastern border contain a long line that's twice bent? And why did New York snip off tiny sections from the southwest corners of Massachusetts and Vermont? How come New York's northernmost border suddenly becomes a straight line instead of continuing on as it had been along the St. Lawrence River?

he land that eventually became New York was previously part of Dutch New Netherlands. (See Figure 114, in NEW JERSEY.) When the British defeated the Dutch in 1674, King Charles II gave a large portion of England's new acquisition to his brother, the Duke of York. The colony of New York assumed that its boundaries were those of the former New Netherlands down to the border of New Jersey. Accordingly, New York continued those border disputes in which the Dutch and their neighboring British colonies had been engaged.

The New York/Connecticut Border

Along almost the entire length of its eastern edge, New York wished to preserve the Connecticut River as its boundary. The Connecticut River, after all, had been the border claimed by the Dutch. The English colonists in Connecticut, however, sought to preserve their ports along the coast of Long Island Sound. In fact, they had already negotiated the Treaty of Hartford with Dutch governor Peter Stuyvesant back in 1650, in which both sides agreed upon a border located 10 miles east of the Hudson River.

But New York pointed out that the Dutch government never ratified the Treaty of Hartford. And instead of a 10-mile buffer east of the Hudson, New York wanted 20. (Figure 121) For Connecticut, such a buffer would have been disastrous, as it would have swallowed up the towns of Greenwich and Stamford.

In 1684, New York and Connecticut commissioned surveyors to devise a boundary that sought to accommodate Connecticut's existing settle-

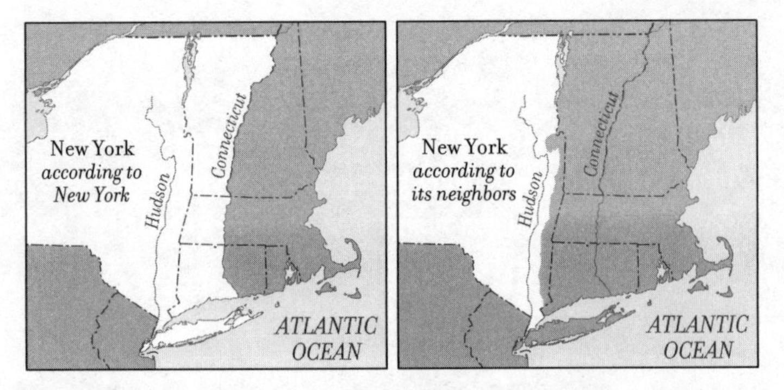

FIG. 121 The Disputed Eastern Border of New York

ments and still achieve New York's objectives. The surveyors came up with a two-part solution. The first part was the creation of a panhandle at the southwest corner of Connecticut, preserving its existing settlements. (See Figure 40, in CONNECTICUT.) Since this panhandle resulted in a loss of land from New York, part two compensated for that loss by releasing to New York an equivalent amount of land farther north. From Ridgefield, Connecticut, to the Massachusetts border, a long slice of land that came to be known as "the Oblong" was given to New York. (See Figure 41, in CONNECTICUT.)

The New York/Massachusetts Border

The middle segment in New York's twice-"bent" eastern border is its boundary with Massachusetts. Since each of the colonies to the east of New York separately negotiated its boundary with New York, the result was a series of separately angled straight-line segments that, taken together, appear to be bent twice.

In the case of Massachusetts also, the dispute emanated from the overlapping boundary claims regarding the Connecticut River (the border claimed by New York) and the Hudson River (the border claimed by Massachusetts). In this instance, however, the two colonies could not resolve their differences, and England intervened. In 1759, the British government declared that the New York/Massachusetts boundary was to be a straight line 20 miles east of, and as parallel as possible to, the Hudson River. To this day, that line remains the New York/Massachusetts border. Almost.

The "almost" regards the southwest corner of Massachusetts. The mountainous terrain at this location is such that, at the time, the only roads to this area's town, Boston Corners, were via Connecticut or New York. This roundabout access was more than inconvenient. Since it interfered with the ability of Massachusetts to maintain order, Boston Corners became a hive of disorderly individuals. Eventually, Massachusetts offered this land to New York, which was only too happy to clean out the hive. In 1855, Congress approved the transfer. So New York did not snip off the corner of Massachusetts. Massachusetts snipped it off itself. (See Figure 93, in MASSACHUSETTS.)

The New York / New Jersey Border

When the Duke of York divided his portion of the New Netherlands, he stipulated that the border between New York and New Jersey was to be a straight line from the source of the Delaware River, "which is in one and forty degrees and forty minutes of latitude," to the point where the Hudson River crosses 41° N latitude. As it turns out, the Delaware reaches nearly to 42°30', over 50 miles farther north than the Duke of York realized. New Jersey never sought to claim this larger boundary, but it did claim its northwest corner at the latitude stated by the duke, 41°40'. In doing so, New Jersey laid claim to a very narrow strip of land along the northern side of the Delaware River. (See Figure 115, in NEW JERSEY.)

New York had always assumed that its border with New Jersey extended along the Delaware River itself down to the point at which the river makes a 90-degree turn. A good deal of unpleasantness and even violence ensued when both colonies sought to collect taxes, record deeds, and otherwise assert their authority over the same narrow strip of land.

After the Revolution, New Jersey agreed to let the river itself be the border. This enabled the two states to argue over New Jersey's eastern border with New York. (For the details on this dispute—and it's a dilly—go to NEW JERSEY.)

The New York / Pennsylvania Border

New York's other neighbor to the west is Pennsylvania—though there was a possibility that it could have been Massachusetts and Connecticut, since their royal charters stated that their western borders were the Pacific Ocean. Both states were blocked, however, by New York. As Indian treaties cleared the land bounded by Pennsylvania, the Catskill Mountains, and Lake Ontario, New York, Massachusetts, Connecticut, and Pennsylvania all asserted their claims in the region. (Figure 122)

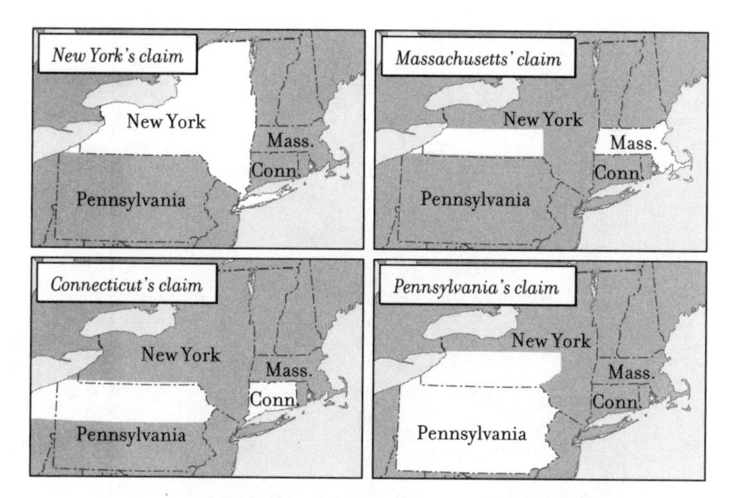

FIG. 122 **Western New York—Overlapping Colonial Claims**

New York was particularly motivated with regard to its claim since, by the end of the Revolution, the idea of a canal linking Lake Erie with the Hudson River was becoming increasingly feasible. Though such a canal's exact route was not yet certain, New York feared that the canal would have to pass below 43°, which Pennsylvania claimed as its northern border. (Figure 123) New York maintained that the northern border of Pennsylvania was 42°, despite the fact that Pennsylvania's royal charter stated that it was the bounded on the north by "the beginning of the three and fortieth degree of northern latitude." (For more on this, go to PENNSYLVANIA.)

In 1785–6, the federal government helped broker a solution among the claims of New York, Massachusetts, Pennsylvania, and Connecticut. The key to the agreement was Virginia.

Virginia (which, at the time, included West Virginia) and Pennsylvania had been engaged in a dispute over the Ohio River region of western Pennsylvania—or Virginia—depending on your point of view. (For more details, go to PENNSYLVANIA.) Once settled, Pennsylvania was more amenable to compromise regarding its northern border's access to Lake Erie. Pennsylvania ceded to New York its claims north of 42°, with the exception of land west of the longitude of Lake Ontario's western end.

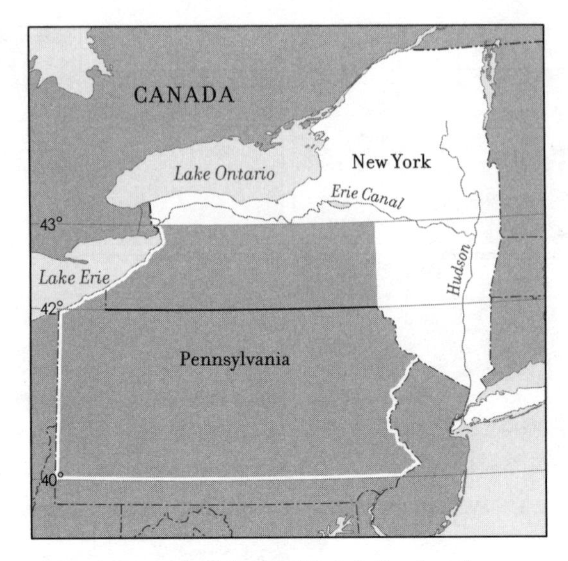

FIG. 123 The Erie Canal in Relation to 43°

Pennsylvania thereby maintained 30 miles of waterfront on a Great Lake that included an excellent port at what is now the city of Erie, and New York obtained the border it wanted to enable it to build the Erie Canal. (Figure 124) (Connecticut, for its part, was given land holdings it claimed farther west, in the area that is now northern Ohio. And Massachusetts obtained the ownership rights to some 6 million acres of land in western New York, though the land remained under the sovereignty of New York.)

The New York/Canadian Border

New York's northern border with Canada dates back to 1763, the year England signed its peace treaty with France ending the French and Indian War. In addition to acquiring the land between the Mississippi and Ohio rivers, England also acquired what is now the province of Quebec, which included its priceless avenue of commerce, the St. Lawrence River.

But England also acquired problems—the foremost of which was its new subjects, who were French and Catholic. Understandably, these Quebecois were concerned about all those English-speaking Protestants

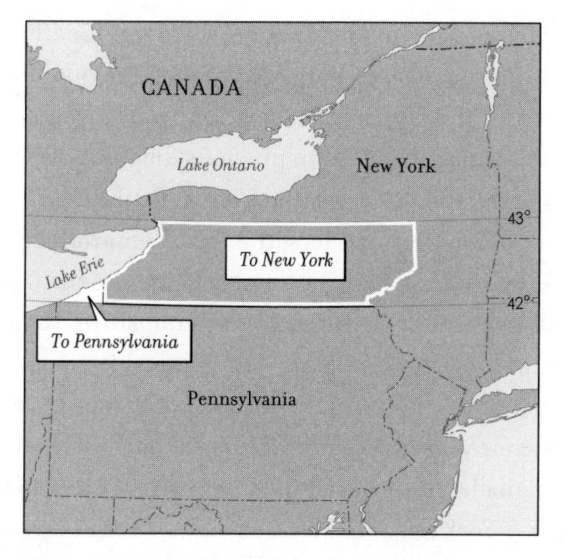

FIG. 124 New York/Pennsylvania Boundary Agreement—1785

who had settled in New England. The Americans were, at the time, furious with England for prohibiting them from expanding into the land they had helped England win west of the Ohio River. Would their eyes now turn north to the regions around the St. Lawrence?

King George III was concerned about American expansionism leading to problems in England's lucrative Canadian commerce, which relied upon the St. Lawrence River. The king therefore created a border to divide the Americans from the Quebecois. The line he defined followed the St. Lawrence until it crossed the 45th parallel. At this point the border became a straight line east to the Connecticut River. The line was designed to provide a buffer for the security of Montreal and to encompass the majority of the French-speaking people. Evidently, it did its job well, for it remains in place to this day. (See Figure 165, in VERMONT.)

The New York / Vermont Border

That segment of American/Canadian boundary that follows the 45th parallel initially resulted in a protrusion on the upper eastern end of

New York. Today, we call this protrusion Vermont. (See Figure 165, in VERMONT.) At the time that George III created the boundary between the Americans and the Quebecois, New York was engaged in a border dispute with New Hampshire. As in previous disputes, New York maintained that its eastern border was the Connecticut River. In those disputes, New York ultimately had to settle for a compromise. But in this instance, the king ruled in favor of New York, despite the fact that it was a far larger and wealthier colony than New Hampshire. In fact, the king ruled in favor of New York *because* New Hampshire was smaller and poorer and, therefore, more inclined to expand into Quebec. Consequently, in the same year that George III set 45° as the boundary between New York and Canada, he declared the Connecticut River to be the eastern border of New York north of Massachusetts. (See Figure 112, in NEW HAMPSHIRE.)

Indeed, New Hampshire was hungrier politically and geographically. Many of those Americans then living on land granted by New Hampshire in the disputed region (grants that were now invalid) joined the ranks of the Green Mountain Boys, under the leadership of Ethan Allen. New York, in turn, sent forces to protect its tax collectors. Before the conflict erupted, however, a larger conflict erupted: the American Revolution.

Vermont continued to seek recognition as a separate state during the Revolution, even going so far as to threaten to ally itself with England. (For more on these maneuvers, go to VERMONT.) In 1791, Congress yielded to Vermont's threats and New York consented to a new border. From the northwest corner of Massachusetts, the border now headed northwestward to a point 20 miles from the Hudson. (So New York did not, as it might appear, snip off the southwest corner of Vermont. Rather, the line angles away from the point where it leaves Massachusetts to readjust its position to the 20-mile margin.) As with New York's borders with Massachusetts and Connecticut, its border with Vermont runs as a straight line, rather than attempting to follow every twist and turn in the Hudson. But in the case of Vermont, the Hudson peters out. In lieu of the Hudson, the straight line continues to the Poultney River, at which point the border follows that river to Lake Champlain, then follows Lake

Champlain northward until reaching 45° N latitude. (See Figure 166, in VERMONT.)

After the Revolution, Congress would locate the nation's internal borders with the goal that all states should be created equal. There would be exceptions to this, most notably Texas and California, but they would be the exceptions that prove the rule. Before the Revolution, however, no such notion of equality prevailed, either in terms of colonial borders or in terms of the people who lived within them. We can see this disparity in the fact that the thirteen American colonies were widely different in size, in fact even more so than they are today.

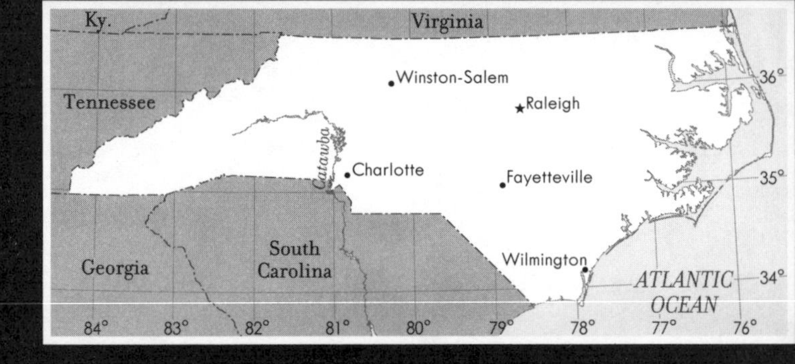

How did North Carolina's border with South Carolina end up being all those angles and steps? Wouldn't the two Carolinas be more equal if they had continued the angled line that begins their division at the coast? Is there some reason, other than convenience, why the northern and southern borders of North Carolina line up with those of Tennessee?

The land out of which North Carolina emerged was part of a grant given in 1629 by King Charles I that created the colony of Carolina. The Carolina Colony consisted of the land between the 31st and 36th parallels, from the Atlantic Ocean to the Pacific Ocean. (See Figure 151, in SOUTH CAROLINA.) As with England's coast-to-coast claims for the colonies of Virginia and Massachusetts, Carolina represented a colossal amount of real estate.

North Carolina's Northern Border

As it turned out, the recipient of the 1629 grant, Sir Robert Heath, never made use of his colony. An identical grant was reissued in 1663 by King Charles II to a group of political allies. Not long after issuing the renewed Carolina charter, the king received a plea urging him to relocate the border farther north. It turned out that 36° passed through the middle of Albemarle Sound and Virginia was imposing its export tax on Carolina's shipping.

Charles II revised the charter, relocating the northern border of the colony at 36°30'. (Figure 125) This latitude passes halfway between

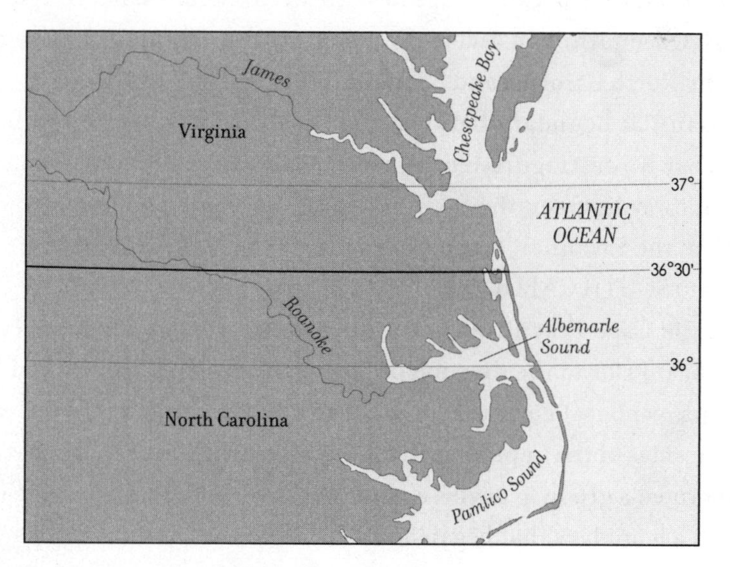

FIG. 125 The North Carolina/Virginia Border

Albemarle Sound and the Chesapeake Bay. As a result, both colonies were endowed with a major waterway that they could employ for commercial purposes.

Though he had no way of knowing it, Charles II had just established a line that would figure prominently in American history—so much so that today we can see it extend west (with a correction here and there) all the way through Missouri and then make a final appearance as the northernmost border of Texas. (See Figure 10, in DON'T SKIP THIS.)

North Carolina's Southern Border

Although this northern border would prove so significant in years to come, back then the Carolina Colony's most important region was farther south at Charleston harbor. From the outset, the distances and differences between the Albemarle Sound colonists and the Charleston colonists created difficulties in governing Carolina as a single colony. Charleston was settled by newcomers and prosperity came to them rather quickly. Around Albemarle Sound, many of the settlers had migrated from Virginia and Pennsylvania and had endured more geographic and political rigors. The two groups never became enemies, but in 1710 they did ask Queen Anne to divide the colony in two. The queen consented, creating North Carolina and South Carolina.

The initial boundary between the Carolinas was deemed to be the Cape Fear River. (Figure 126) It had the advantage of being almost exactly midway between the two Carolinas, the southern border of which was then the Savannah River. (For more on the many shifts in this border, go to SOUTH CAROLINA.)

But the Cape Fear River never did become the border between the Carolinas. Prior to the division of the colony, North Carolina had been given its own local government and had already granted tracts of lands on both sides of the Cape Fear River. Consequently, in 1730 King George II appointed a group to devise a more workable border. These men came up with a boundary that began 30 miles down the coast from the mouth

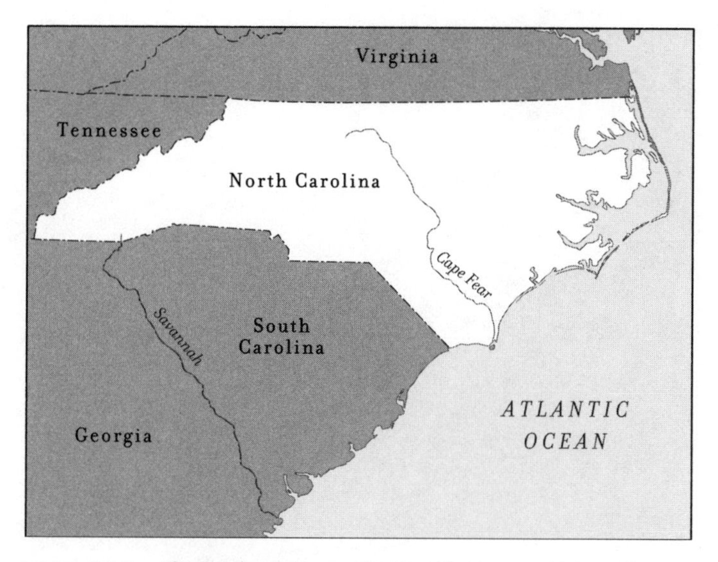

FIG. 126 The Border Between the Carolinas—1710 Proposal

of the Cape Fear River. From that point, the line proceeded northwest to the 35th parallel, then due west to the Pacific Ocean. (Figure 127)

As it turned out, such a line would have sliced through the Catawba Indians' land. The British wanted to avoid conflict with the Catawba and, to their west, the Cherokee, since both tribes were allied with England against France and Spain. So the king's appointees unfurled yet another border in 1735. This time, adjustments were to be made in the line when it encountered the Catawba's land. Amazingly, however, it did not intersect their property. The reason was that the line was surveyed 13 miles south of the *actual* 35th parallel (Figure 128)

When the mistake was later discovered, the British government decided in 1771 to compensate South Carolina for the 800 square miles the mistake had cost them, by adjusting the border farther west. The boundary, then, would remain as is up to the Catawba lands, at which point it would continue around the northern borders of the Catawba lands to the Catawba River, then follow the Catawba River to the point where it branches and from there head due west. This westward line would compensate South Carolina, since it would be north of 35°. A little extra

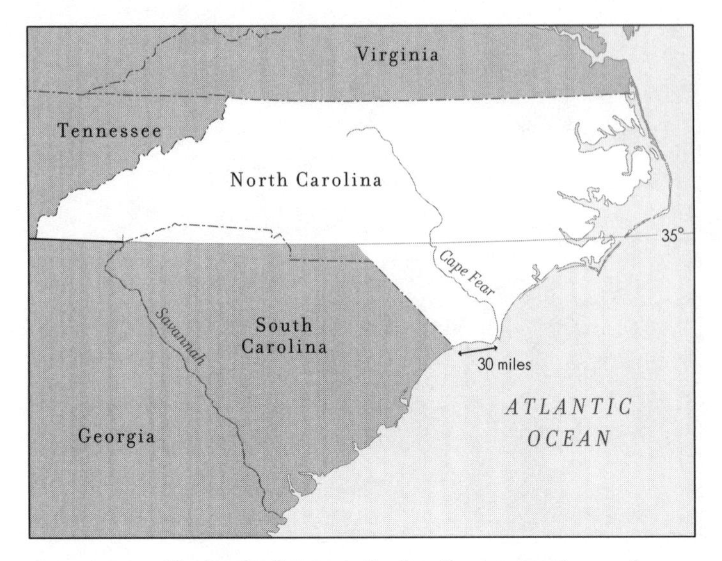

FIG. 127 The Border Between the Carolinas—1730 Proposal

compensation resulted from the fact that, this time around, the surveyors mistakenly veered northward as their line proceeded west. (Figure 129) This ruling explains why the North Carolina/South Carolina border has its angles and steps, and even that odd little notch.

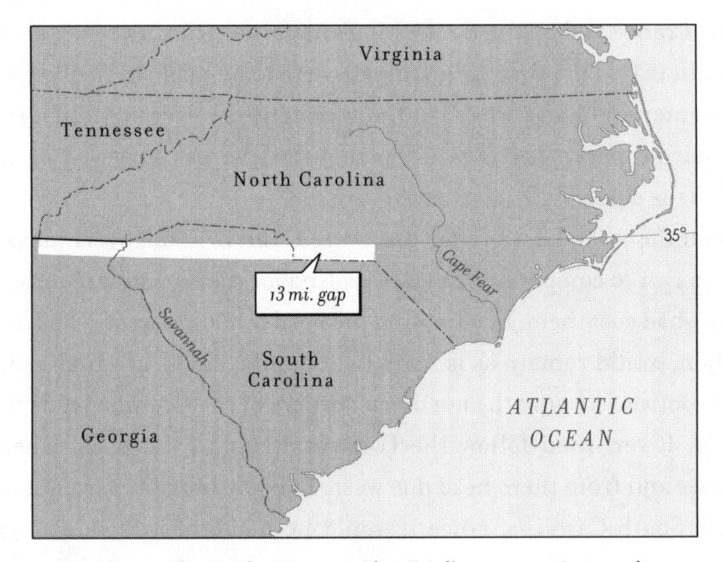

FIG. 128 The Border Between the Carolinas—1735 Proposal

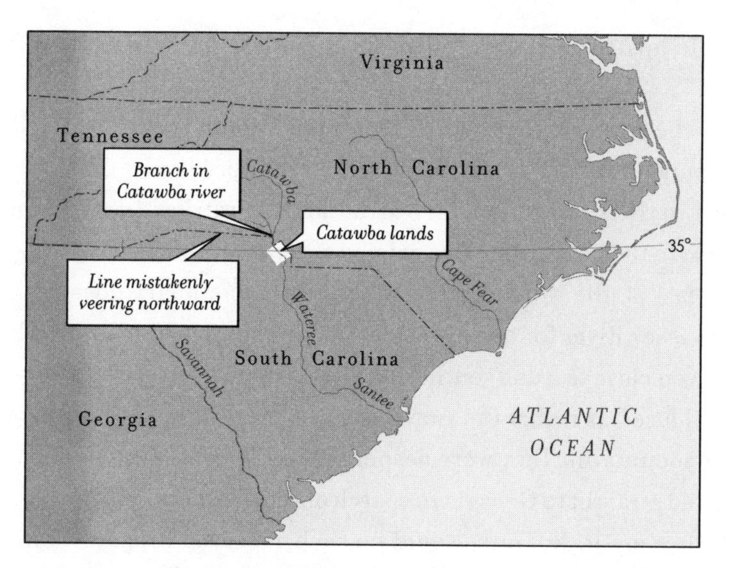

FIG. 129 **The Border Between the Carolinas—1771 Update**

North Carolina's Western Border

The 1783 Treaty of Paris, which ended the Revolution, stipulated the boundaries of the newly born United States. The western boundary of North Carolina was defined as being the Mississippi River. At that time, the United States consisted of former colonies that were widely different in size, particularly those with borders that extended beyond the Appalachian Mountains. The federal government successfully convinced these states to donate their lands west of the mountains to the United States. By doing so, more states could be created and all the states would be more equal.

The land that North Carolina released became Tennessee. For the dividing line, North Carolina called for the boundary to follow, as closely as possible, the crest of the Appalachians all the way down to the Georgia border.

But this is not the boundary the surveyors marked. At the southwest corner of the state, they departed from the crest of the mountains at the point where it intersected the Unicoi Turnpike, and headed south to the Hiwassee River. From here, according to their report, they proceeded one mile upstream then headed due south all the way to the Georgia border.

Why the surveyors, who were appointed by both states, deviated from the stipulated border is a mystery. Also a mystery is why the border described in their report (which North Carolina ratified in 1821) does not conform to the border that has existed since that time!

The fact is, the North Carolina/Tennessee boundary does not follow the Hiwassee River for one mile, and the line south to the Georgia border is not even close to a due south line. Why? (Figure 130)

Local lore has it that the surveyors had spent so many months in the remote mountains, they were desperate for the comforts of a tavern and so headed straight to Georgia in search of refreshment. This explanation is clearly a myth. Not only would no such surveyed line ever have been accepted, but the fact is that the surveyors were working in the heart of moonshine country (as some of their journal entries verify).

Another theory is based on the fact that the boundary stipulated on paper was ambiguous at points. Indeed, choices were made by the surveyors farther north that resulted in an advantage going to North Carolina. According to this theory, the departure from the crest of the

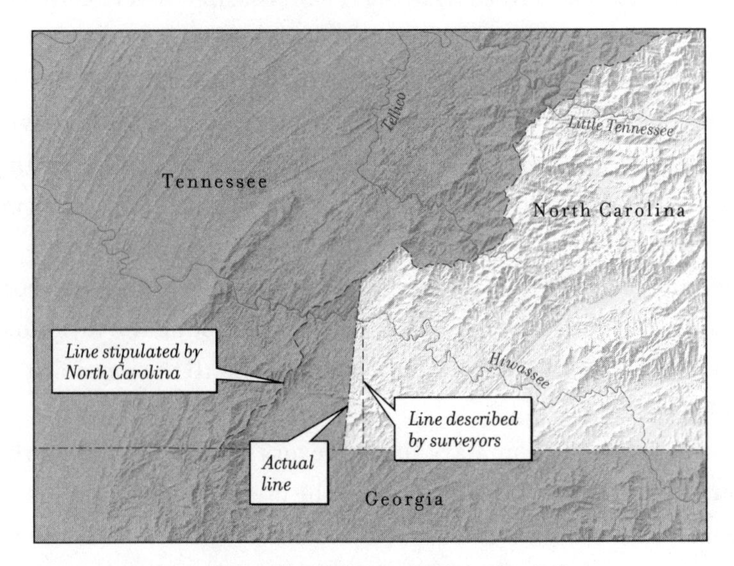

FIG. 130 North Carolina/Tennessee Border Deviation

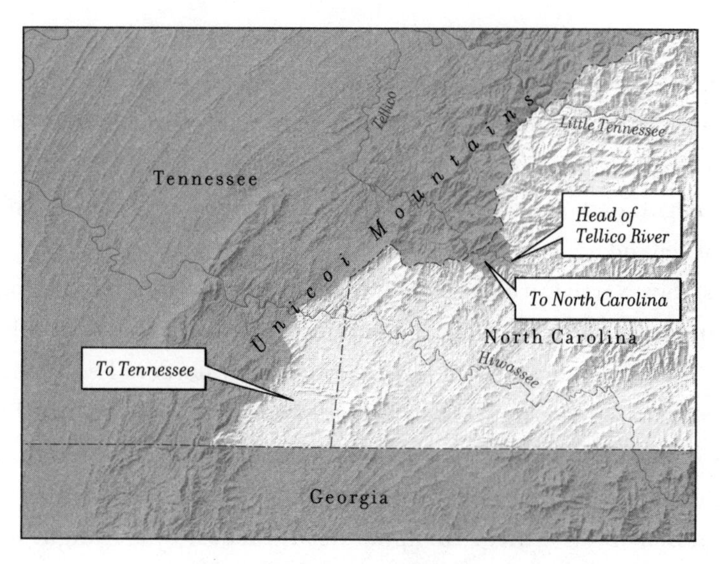

FIG. 131 North Carolina/Tennessee Border—Trade-off?

mountains and straight line south were compensation for the choice made farther north. (Figure 131)

But this theory, too, is dubious. If there was a trade-off, it was a very lopsided deal, greatly favoring the offspring state, Tennessee, over the state that created it. The report of the surveyors made no reference to a trade-off. And had there been one, it is unlikely North Carolina would have later disputed the boundary, which it did—but not the segment that caused the large loss of land by departing from the mountain crest.

Several clues, however, suggest a different deal may have been made.

To fight the Revolution, the newly born states incurred huge amounts of debt. North Carolina sought to reduce its debt by selling parcels of its colonial-era region west of the mountains. In addition, it sold warrants for Cherokee land that could be exchanged for title to that land at such time as the Cherokee released it. (The fact that the days of the Cherokee in the South were numbered was, evidently, common knowledge.)

In 1784, North Carolina agreed to cede it western region to enable the creation of a new state, which became Tennessee. But until Tennessee acquired statehood in 1796, North Carolina continued selling warrants and titles to its land. Tennessee disputed these sales. As this and other conflicts approached the point of violence, the federal government intervened. The first clue regarding the mysterious southwest boundary of North Carolina appears in the 1806 agreement that the federal government negotiated. It voided the warrants North Carolina had sold for Cherokee land.

Possibly, North Carolina sought to offset any liability it might have in pending legal battles (regarding duplicate sales and fraud) by allowing the straight line departure from its stipulated border. The location of this departure—clue number two—just happens to have occurred at the point where the recent 1819 treaty with the Cherokee defined this segment of their reservation as proceeding along the crest of the Unicoi Mountains "to the Unicoi Turnpike Road; thence, by a straight line, to the nearest main source of the Chestatee."

Neither the line described by the surveyors nor the actual line proceeds toward the nearest source of the Chestatee. By proceeding south, however—clue number three—both lines divide nearly equally that part

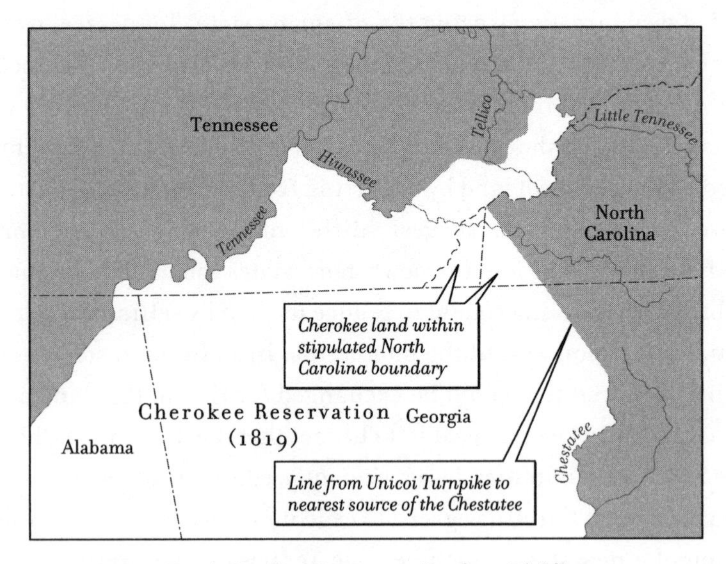

FIG. 132 North Carolina/Tennessee Border—Dividing Cherokee Land?

of the Cherokee land that would have been in North Carolina had the surveyors followed the originally stipulated border. (Figure 132)

Perhaps North Carolina agreed to divide this land that the federal government had cleared of dubious sales in return for Tennessee not seeking compensation from North Carolina for its remaining dubious sales.

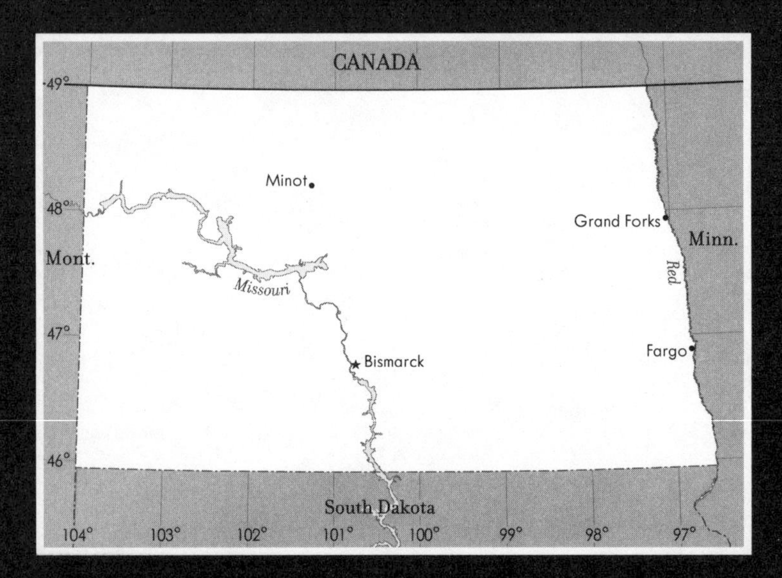

Why is the straight-line western border of North Dakota located where it is? Is North Dakota's southern border located where it is to divide the two Dakotas in half? If so, how come North and South Dakota are not quite equal in height?

North Dakota's Northern Border

The area that is today known as North Dakota was mostly, maybe entirely, acquired by the United States in the 1803 Louisiana Purchase

(See Figure 5, in DON'T SKIP THIS.) The precise boundaries of the land France sold to the United States were not explicit. Consequently, England and the United States agreed in 1818 to extend the border between Canada and what is now Minnesota westward to the crest of the Rocky Mountains. (To learn why they chose the 49th parallel in the first place, see DON'T SKIP THIS.) This agreement created the northern border of North Dakota.

North Dakota's Eastern and Western Borders

When Minnesota became a state in 1858, it shed most of its territorial land and acquired a new western border. Three years later, Congress created the Dakota Territory. (Figure 133) Minnesota's western border became the eastern border of the Dakota Territory and, in time, part of it, the Red River of the North, became the eastern border of North Dakota. (For details as to why that border is located where it is, go to MINNESOTA.)

The western border of North Dakota surfaced as a result of Idaho, despite the fact that Idaho, as we know it today, shares no border with North Dakota! In 1863, Congress created the Idaho Territory in response to gold discoveries in the region's mountains. Initially the territory included all of present-day Idaho and nearly all of the land east of Idaho to the 104th meridian. (Figure 134) This created a new western border for

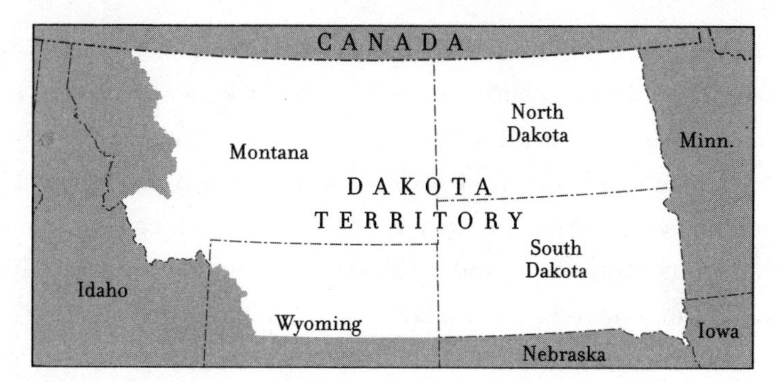

FIG. 133 Dakota Territory—1861

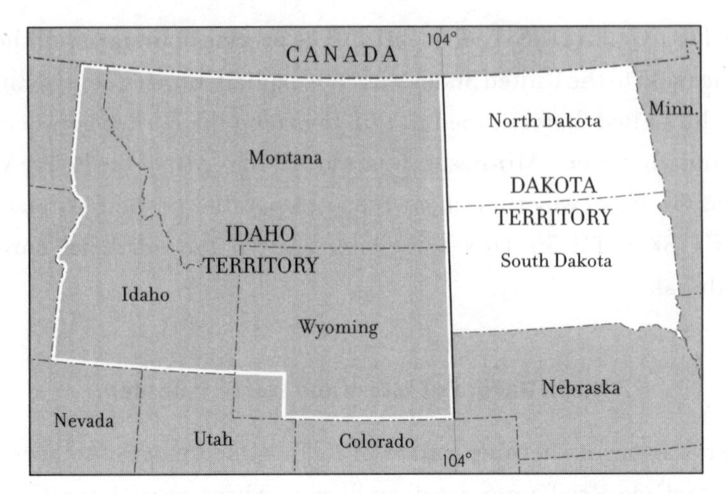

FIG. 134 The Emergence of North and South Dakota's Western Border—1863

the Dakota Territory, one that remains to this day as the western border of North and South Dakota.

But why was this border located at 104°? This border is an artifact of the effort by Congress to create states that are as equal as possible. By locating the western border of the Dakotas at the 104th meridian, North Dakota joined South Dakota, Wyoming, Colorado, Washington, and Oregon in having almost exactly seven degrees of width. (See Figure 13, in DON'T SKIP THIS.)

North Dakota's Southern Border

In preparation for statehood, the Dakotas divided into North and South Dakota in 1887. The legislation called for the Dakota Territory to be divided along the 46th parallel, which bisected the Dakota Territory's six degrees of height (43° to 49°). In effect, the location of this line had been determined over thirty years earlier, when Congress created Kansas and established its southern boundary at 37°. In doing so, Congress altered, by one-half of a degree, a boundary line that extended from Virginia/North Carolina, through Kentucky/Tennessee, and Missouri/Arkansas. Why make this change? By locating the southern border of Kansas at 37°,

Congress could eventually create a column of four states, each having three degrees of height. And in fact, Kansas, Nebraska, South Dakota, and North Dakota all possess three degrees of height. Almost.

In actuality, the southern border of North Dakota is located not at 46° but at 45°55'—one-twelfth of one degree off. Surveying error? In this instance, no. The Coteau des Prairies is a 200-mile-long plateau that stands out above the prairies. The northern tip of this extraordinary landform is just south of the 46th parallel, but it provides an inviting location for the border. And indeed the boundary is located just at the tip of the Coteau des Prairies. (See Figure 153, in SOUTH DAKOTA.)

Although the tip of the Coteau des Prairies is only one-twelfth of a degree south of the 46th parallel, to some extent the land South Dakota lost compensates North Dakota, which is slightly smaller because its eastern border, the Red River of the North, veers slightly westward.

This adjustment to the North Dakota/South Dakota border because of the Coteau des Prairies serves two very American values: practicality and equality. Indeed, those two values are reflected not only in the borders of North Dakota but in many of the similarly shaped states that are characteristic of the American west—states whose borders were never tainted by colonial rule.

OHIO

Why are Ohio's straight-line borders located where they are? And why is Ohio's northern border almost, but not quite, aligned with that of its neighbor, Indiana?

The land that is now Ohio was previously part of the Northwest Territory, the land acquired by the British and their American colonists from France in the French and Indian War (1754–63). (See Figure 4, in DON'T SKIP THIS.) In 1787, Congress enacted the Northwest Ordinance, which specified the lines that would someday subdivide the Northwest Territory into states.

Ohio's Western Border

In 1800, Congress began to execute its division of the Northwest Territory by establishing a vertical border that separated the eastern region (the Ohio Territory), where American settlements were increasingly being established, from the Indian lands to the west (the Indiana Territory). This vertical line proceeded north from the juncture of the Ohio River and the Great Miami River. (Figure 135) While the original line went all the way to the Canadian border, its remnant can still be seen today as the western border of Ohio.

Ohio's Northern Border

Congress further divided the Ohio Territory two years later in preparation for Ohio's becoming a state. This time, however, things didn't go quite so

FIG. 135 Ohio's Western Border—1800

smoothly. The surveyors discovered that a line due east from the southernmost point of Lake Michigan, as called for in the Northwest Ordinance, would separate Ohio from the point where one of its key western rivers, the Maumee, empties into Lake Erie. Today we call that point Toledo.

Congress accordingly revised the line defining Ohio's northern border. It would now commence just above Toledo and from there head straight for the southernmost point of Lake Michigan, stopping when it reached Ohio's western border. (Figure 136)

Michigan, however, was not pleased, as this adjustment cut into some of its most valuable land. As the disagreement became heated, Michigan dispatched its militia to seize the land in question. Ohio met this challenge, and in 1835 the Toledo War, as it became known, erupted. To ease tensions, Congress broke off a chunk of what would have been the upper peninsula of Wisconsin (hardly any settlers were living there yet to complain) and gave it to Michigan as compensation.

Because Ohio's northern border was adjusted, and because Indiana, in due course, also adjusted its northern border to obtain frontage on Lake Michigan, what had been intended to be a continuous line defining the northern borders of both states is today separate segments that are not quite aligned.

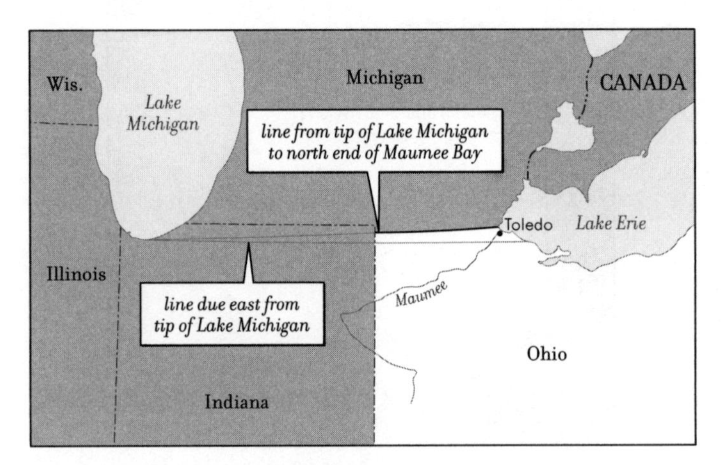

FIG. 136 Ohio's Adjusted Northern Border

Ohio's Eastern Border

Ohio's eastern border also had a scrappy past. Pennsylvania, Virginia (which at the time included West Virginia), and, of all places, Connecticut each laid claim to overlapping areas along what today is the eastern border of Ohio. Evidence of this multistate brawl can still be seen in the finger of West Virginia wedged between Ohio and Pennsylvania. (Figure 137) How did that happen?

Following the French and Indian War, England forbade the Americans to migrate beyond the Ohio River, despite the fact that the colonists had fought side by side with the British to oust the French from these regions. England's prohibition outraged the Americans. To mollify the colonists, England granted permission to Virginia to form corporations for investing in land along the Ohio River.

Connecticut's claims in what is now Ohio were based on its 1662 charter, which granted it all the land between its northern and southern borders from the Atlantic Ocean to the Pacific Ocean (though subsequent land grants creating Rhode Island and New York had lopped off chunks

FIG. 137 The "Finger" of West Virginia Wedged Between Ohio and Pennsylvania

of its coast-to-coast claims). Like Virginia, Connecticut also received permission from England to form corporations to invest in its western regions. (Figure 138)

Ohio's eastern border first began to surface when Virginia and Pennsylvania settled their dispute over Virginia's Ohio River land claims in relation to Pennsylvania's western border. The line Virginia and Pennsylvania negotiated enabled Virginia to keep its investments along the Ohio River, but created the narrow finger of land that remains to day. As part of the agreement, Virginia released its claim to any land beyond the Ohio River. As a result, the Ohio River became the eastern border of Ohio up to the point where it crossed the newly defined western border of Pennsylvania, which then takes over as Ohio's eastern border.

But one hurdle remained. Connecticut. It maintained its western claims, which included the northern tier of what was hoping to be Ohio. Massachusetts, Virginia, Georgia, and North Carolina were all releasing their extended land claims to enable the creation of new states. Connecticut, however, being so much smaller, released all but

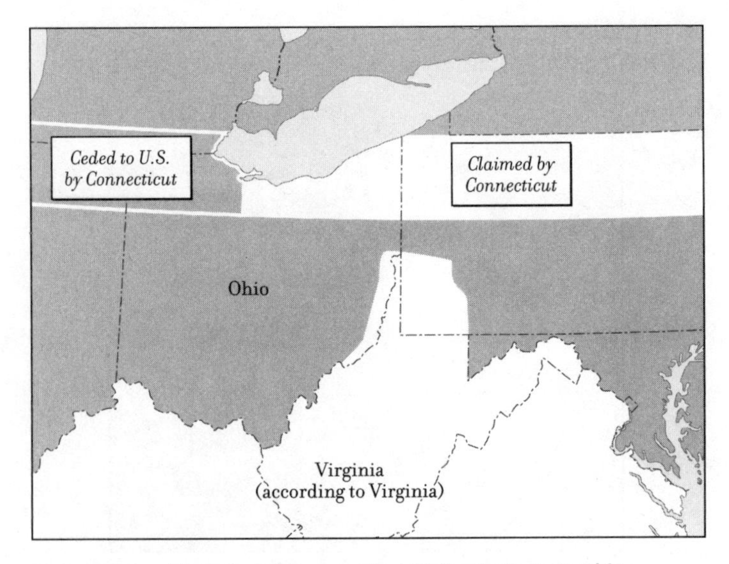

FIG. 138 Virginia and Connecticut Claims in Eastern Ohio

the easternmost section of its western claims, a chunk it called its Western Reserve. Finally, in 1800, Connecticut threw in the towel and released this land, too. To this day, however, a remnant of Connecticut's presence remains, embedded in the name of a university located in Cleveland, Case Western Reserve.

OKLAHOMA

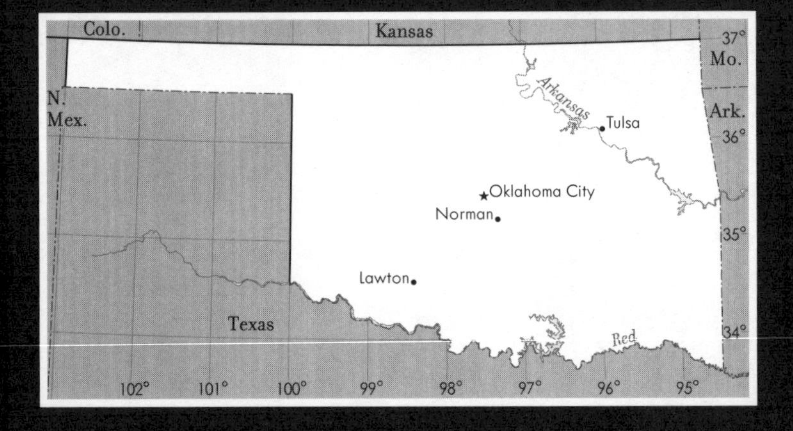

Why in the world does Oklahoma have that skinny panhandle? Why didn't Texas just fill up that space? And why is Oklahoma's straight-line eastern border bent twice? How come the straight-line northern border of Oklahoma lines up with the straight-line border between its western neighbors, Colorado and New Mexico, but doesn't quite line up with the straight-line border between its eastern neighbors, Missouri and Arkansas?

Most of what we now call Oklahoma came into American possession with the Louisiana Purchase (1803). Since the boundaries of the land France sold us had not been explicitly stated, the United States entered into negotiations with Spain to define the western extent of the Purchase, and with England to define its northern extent. (See Figure 8, in DON'T SKIP THIS.)

Oklahoma's Southern and Western Borders

As a result of the negotiations with Spain, the first outlines of what would later become Oklahoma emerged in the 1819 Adams-Onis Treaty. One segment of the border dividing the United States from Spain's North American possessions is described in the treaty as following the Red River to 100° W longitude, then due north, to the Arkansas River. (Figure 139) Long before Oklahoma existed as a separate entity, its southern and western borders (not including its western panhandle) surfaced by virtue of one of the first treaties negotiated by the young American government. (To find out why the United States and Spain settled on the

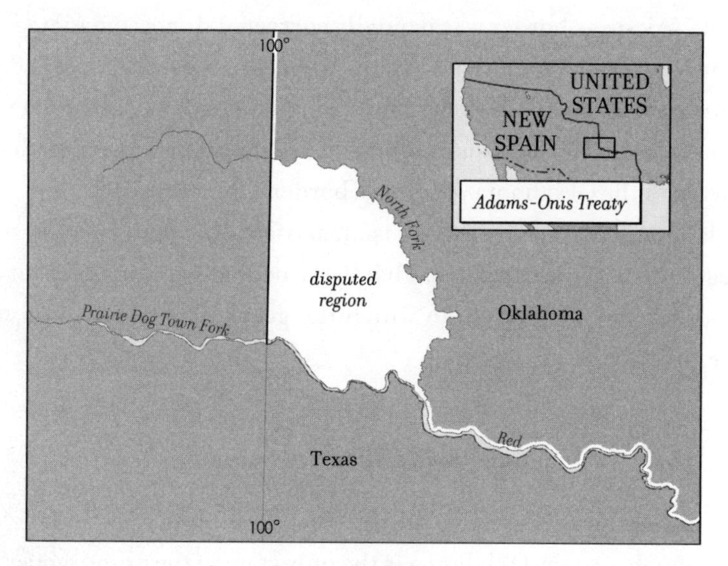

FIG. 139 Southwest Corner of Oklahoma Surfacing in Adams-Onis Treaty

100th meridian as the point at which the border turned due north from the Red River, go to TEXAS.)

Oklahoma's Eastern Border

In 1821, Congress created the territory of Arkansas. Originally, the western border of Arkansas was intended to be a straight-line continuation of the western border of Missouri. This would have resulted in Oklahoma sharing this straight-line continuation as its eastern border. But only a small segment of Oklahoma's eastern border matches the western border of Missouri, that being the portion directly between Oklahoma and Missouri.

The remainder of Oklahoma's eastern border consists of two other segments. These segments resulted from the Treaty of Doak Stand, the document responsible for the removal of the Choctaw Indians from Mississippi to what would become Oklahoma. In negotiating the treaty, the United States inadvertently gave the Choctaw not only land in the future Oklahoma but a hefty slice of Arkansas, too. After renegotiating the treaty (a process that entailed the demise of noncooperative Indian leaders), the Arkansas border was partially corrected. It now emanated from a point 100 paces west of Fort Smith, from which point it proceeded in two directions: due south to the Red River and north by northwest to the southwest corner of Missouri. These two directions are today the two segments of the Oklahoma/Arkansas border. (Figure 140)

Oklahoma's eastern boundary is an artifact of Indian treaties and of "renegotiated" Indian treaties. That this line is twice bent makes it a fitting border to a territory into which the government herded so many Native Americans.

Oklahoma's Northern Border

Oklahoma is the only state in the union to have borders at both 37° and 36°30', which is to say Oklahoma is the only state in the union whose borders embody artifacts of an important shift in American history.

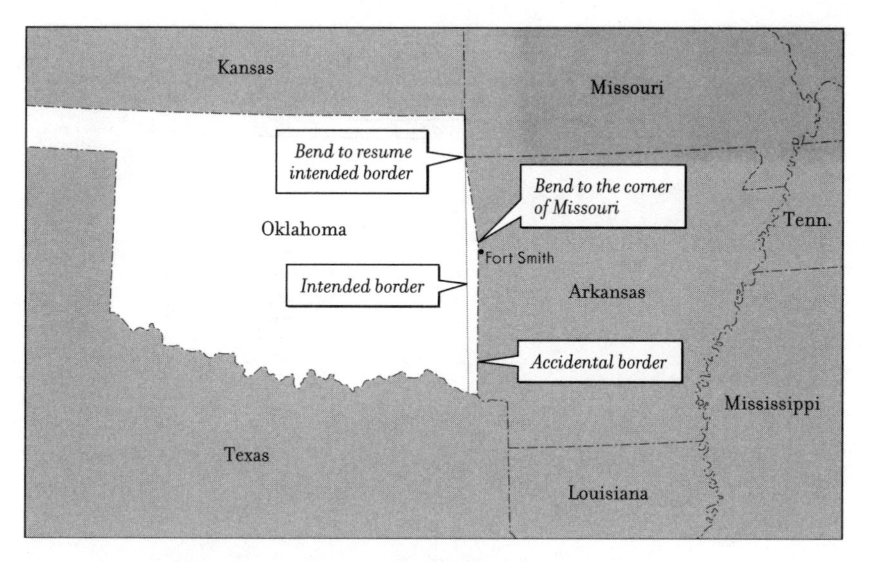

FIG. 140 Oklahoma's Twice-Bent Eastern Border

When the future Oklahoma's neighbor to the south, Texas, entered the Union in 1846, it wanted to maintain slavery, which it had permitted during its days as an independent republic. Under the Missouri Compromise (1820), however, Texas could not be a slave state if its borders extended north of 36°30'. So Texas lopped off its lands north of 36°30' and gave them to the United States. In doing so, Texas created what would later become the southern border of the Oklahoma Panhandle.

Eight years later, in 1854, the Kansas-Nebraska Act relegated the Missouri Compromise to the junk heap of efforts to negotiate slavery. In dispensing with 36°30' as the determinant for where slavery could exist, Congress altered the proposed southern border of Kansas, setting it at 37°. This line would later become the northern border of Oklahoma. (Figure 141) It also enabled Congress, in the years ahead, to create a tier of four prairie states that each had three degrees of height and a tier of three Rocky Mountain states that each had four degrees of height. (See Figures 11 and 12, in DON'T SKIP THIS.)

These two borders, both of which remain today as the northern and southern borders of Oklahoma's panhandle, bear witness to a key shift

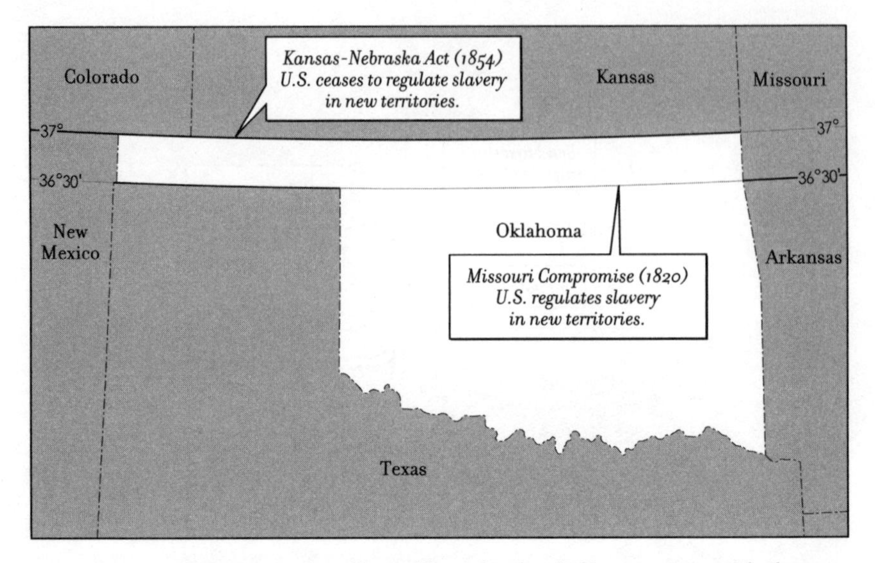

FIG. 141 Oklahoma Panhandle—Failure of Federal Efforts to Cope with Slavery

in the attitudes of the time. Congress was trying to turn its eyes away from the fundamental inequality of slavery (by giving the choice to the states) and fix its gaze on an idealized (indeed, mathematical) vision of equality among the newly forming western states. All this is preserved in the borders of Oklahoma's panhandle, in one-half of one degree of latitude.

Oklahoma's Southern Border Revisited

In 1890, Congress created the territory of Oklahoma. Among the first actions Oklahoma took was to challenge Texas over which branch of the Red River constituted their border under the Adams-Onis treaty. Texas, as one would expect, claimed that the northern branch was the border. (Figure 139) Oklahoma claimed that the southern branch, being larger, was the main branch. In 1897, the U.S. Supreme Court ruled in favor of Oklahoma and this segment of its southern border was altered, giving Oklahoma the shape it has today.

OREGON

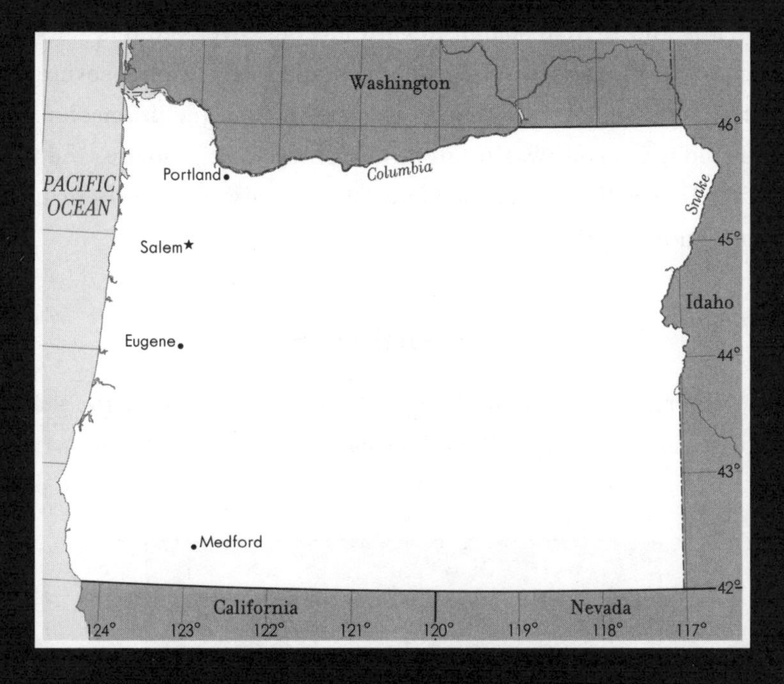

Why is Oregon's eastern border located where it is? And why does it both squiggle and go straight? Why not stick with one or the other? And why is Oregon's straight-line southern border located where it is?

In the 1700s, British fur traders in northern regions between the Pacific coast and the Rockies came into conflict with Russian fur traders arriving from the north and Spanish traders from the south. (Remnants of Spain's presence along the Canadian coast can still be seen in the names of many of the communities and inlets, such as Juan de Fuca Strait, Anacortes, and Lopez Island.) Americans began appearing in the mix in the early 1800s, following the Louisiana Purchase (1803) and the Lewis and Clark Expedition (1804–06).

By the time the Americans arrived, England had already negotiated a boundary agreement with Spain, but not yet with the Russians. The Americans and British sought to gain leverage over the Russians, and eliminate conflict between themselves, by agreeing to joint-sovereignty over the area, which they called the Oregon Country. It encompassed what is today Oregon, Washington, British Columbia, Idaho, and those parts of Alberta, Wyoming, and Montana that are west of the Continental Divide. (Figure 142)

Oregon's Southern Border

The southern border of the Oregon Country—and, later, the state of Oregon—was the result of England's earlier negotiation with Spain. In

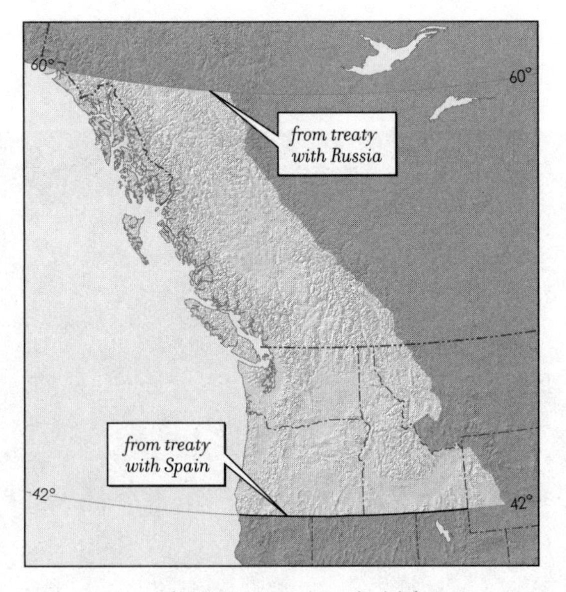

FIG. 142 **The Oregon Country (British & American)—1783–1846**

1790, the two powers agreed that England could have free rein north of the 42nd parallel, which is today the southern border of Oregon.

But why 42°? When these agreements were negotiated, rivers were the avenues of commerce. North of 42°, virtually all the waterways flow into the Columbia River, which empties into the Pacific Ocean at what is today Portland, Oregon. Below 42°, virtually all the waterways flow to San Francisco Bay or directly to the Pacific. The 42nd parallel therefore served as an excellent dividing line. Even now it serves as the southern border of Oregon and Idaho and the northern border of California, Nevada, and much of Utah. (See Figure 9, in DON'T SKIP THIS.)

Oregon's Northern Border

By the early 1840s, both England and the United States were ready to separate their interests in the Oregon Country. England suggested that the dividing line should be an extension of the U.S./Canadian boundary to the east, 49°. The United States claimed it should be the line it had previously negotiated with the Russians: 54°40' N latitude. (For more on this earlier negotiation with Russia, go to ALASKA.) A popular slogan from the 1844 presidential election was "54/40 or Fight!"

England, as it turned out, was fully prepared to fight. A border at 54°40' would have deprived the British of Vancouver, their major Pacific port. President James Polk secured his place as one of America's least remembered presidents by avoiding that war. Though an expansionist at heart, Polk thought with his head and agreed to a continuation of 49°.

The Americans now called their half of the Oregon Country the Oregon Territory. And in 1853, the United States divided the Oregon Territory to create the Washington Territory. The new northern border for the Oregon Territory was the Columbia River to the point where the river first crosses the 46th parallel. The boundary then followed the 46th parallel due east to the Continental Divide. A large portion of this boundary has since remained the northern border of Oregon. (Figure 143)

But if Congress was so committed to equality in the creation of states, why is Oregon so much larger than Washington? Since Oregon's

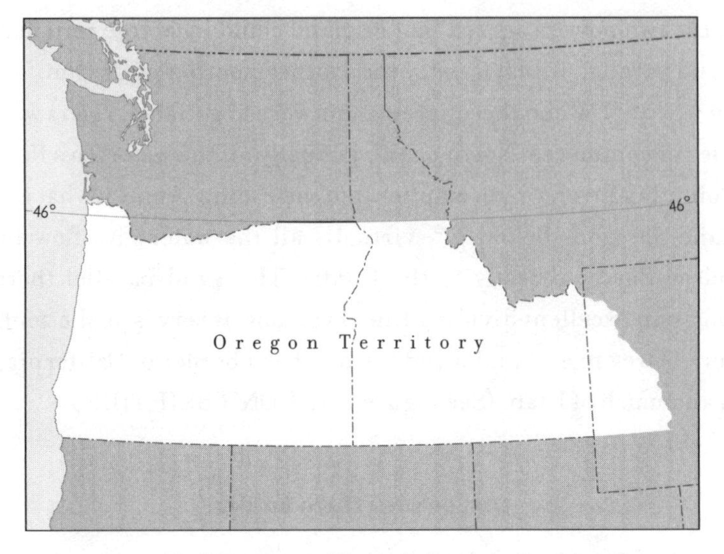

FIG. 143 **The Emergence of Oregon's Northern Border—1853**

southern border is at 42° and Washington's northern border is at 49°, why didn't Congress give both states 3.5 degrees of height? Such a division would have placed the vitally important Columbia River entirely in Washington, giving that state two ports on the Pacific—Seattle and Portland—while Oregon would have had none. The Columbia River ultimately provided a much more equal dividing line.

Oregon's Eastern Border

When Oregon became a state in 1859, its eastern border was modified when it released its eastern territory. Rather than beginning at the summit of the Rockies, the border was now located at the Snake River. It began at the point where the Snake River crosses the 46th parallel, and followed the river south to the point where it is joined by the Owyhee River. From this juncture, a straight line takes over heading due south to the 42nd parallel. (Figure 144)

But why did Congress choose the juncture of the Owyhee River and the Snake River as the point from which the straight line south should

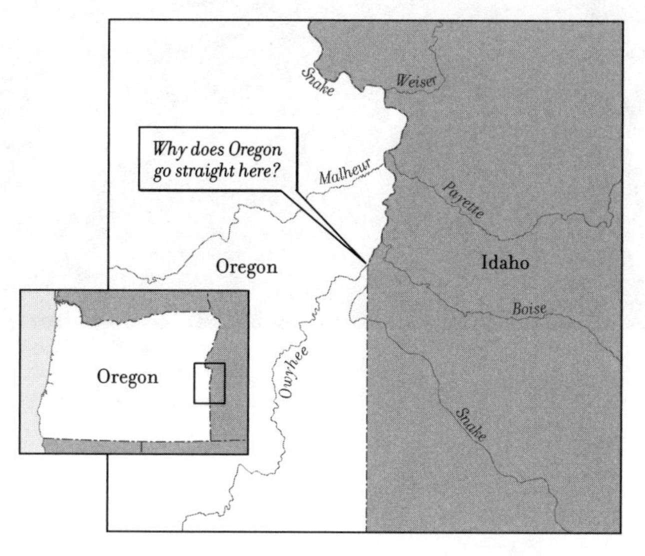

FIG. 144 The Point Where Oregon's Eastern Border Goes Straight

commence? Plenty of other rivers intersect the Snake in this region. Why not use the Burnt River or the Weiser River or the Payette River or the Malheur River or the Boise River?

Thirty years after the establishment of this border, another new border revealed the reason Congress selected the Owyhee. When Washington acquired statehood in 1889, its eastern border was also formed by the Snake River until its juncture with the Clearwater River, at which point a straight line took over. (See Figure 171, in WASHINGTON.) The Snake River's juncture with the Clearwater and its juncture with the Owyhee are both located at the 117th meridian. And the 117th meridian resulted in both Oregon and Washington having almost exactly seven degrees of width. As such, they join four other western states—Wyoming, Colorado, North Dakota, and South Dakota—that have seven degrees of width.

Oregon's borders very much preserve the American quest for equality, though its northern border, the Columbia River, reflects the fact that, when you get down to earth, equality can become a complex question.

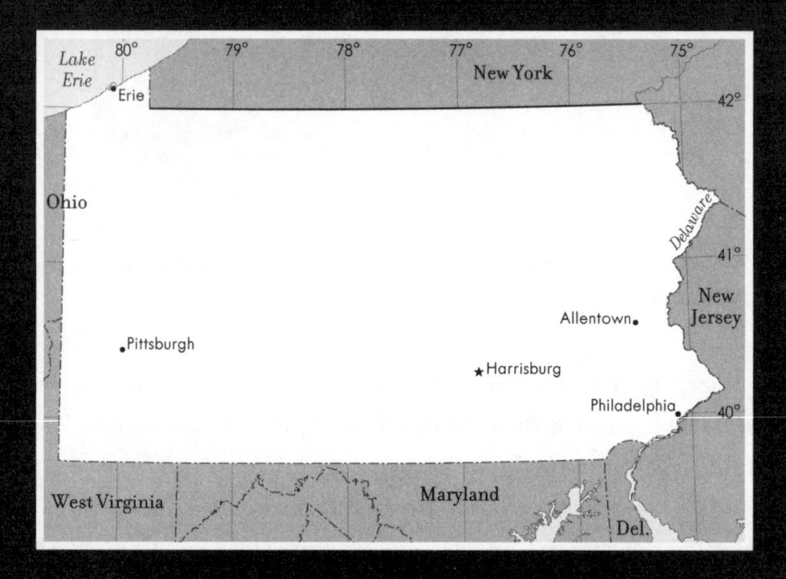

How did Pennsylvania come to get that tab on its northwest corner? And why are its straight-line borders located where they are? How is it that a state as big and as old as Pennsylvania is so close to the ocean without having any frontage on the ocean?

The land that is today called Pennsylvania was first defined as such in 1681, when King Charles II granted it to William Penn as payment for debts owed by the monarchy to the Penn family. Its boundaries, under

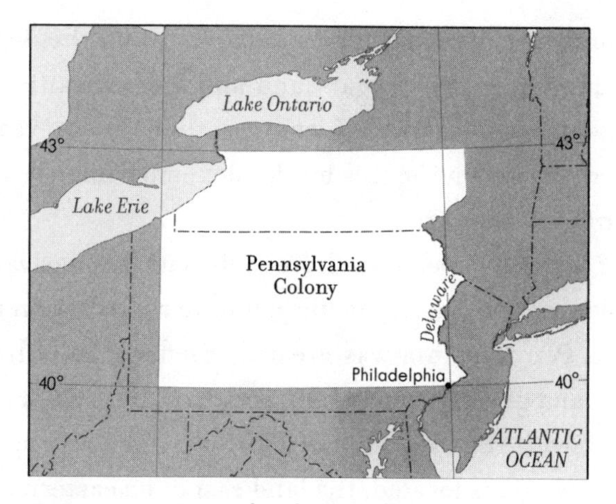

FIG. 145 **Pennsylvania According to the 1681 Charter**

the terms of the charter, were basically those of a rectangle. It extended from the 40th parallel on the south to the 43rd parallel on the north, and from the Delaware River on the east to a straight-line western border five degrees distant. The only unusual element was a semicircle located at Pennsylvania's southeast corner, maintaining a buffer of 12 miles from the Dutch settlement at New Castle. It sounded like a great border. It wasn't. (Figure 145)

Viewing the boundaries of Pennsylvania's charter on a map might lead one to believe that the primary conflict would be with New York, where the 43rd parallel cuts deeply into what is today western New York. But it wasn't western New York then. It was Iroquoi land. Pennsylvania's most intense boundary disputes were with Virginia and Connecticut.

Virginia and Connecticut? They don't even border Pennsylvania! How did this come about? It all began with Maryland.

Pennsylvania's Southern and Eastern Borders

When King Charles I issued Maryland's charter in 1632, he located its northern border at 40° N latitude, which was why Pennsylvania's charter located its southern border at the same latitude. Unfortunately, 40°

turned out to pass right through Philadelphia. Had this been the only point of contention between Maryland and Pennsylvania, they would likely have found a solution in far less time than the 100 years in which they wrangled. But resolving this border depended upon resolving their dispute regarding Delaware.

Delaware? What did Delaware have to do with Pennsylvania? Unlike the other colonies, Pennsylvania did not have a window on the Atlantic Ocean. When Pennsylvania was created, England had already issued charters and land grants covering the entire Atlantic coast of the future United States, with the exception of Florida, which was held by Spain. Where Pennsylvania is located, the land east of its eastern border along the Delaware River had previously been Dutch territory. England ousted the Dutch in 1674, after which King Charles II granted this land to his brother, the Duke of York, under whom it came to be known as New Jersey.

For Pennsylvania, this meant that the Delaware River was more than just its eastern border, it was also its lifeline to the sea. And that is why Delaware was critical to Pennsylvania. The Delaware River flows alongside Philadelphia, then farther south widens into Delaware Bay, which then flows into the sea. Where it widens into Delaware Bay was where the Dutch colony, now known as the state of Delaware, begins. And it was where Pennsylvania's access to the ocean was vulnerable. (See Figure 44, in DELAWARE.)

William Penn sought proprietorship over Delaware. But Maryland objected, pointing out that the region had been included in Maryland's charter. Even though England rejected Maryland's claim, the dispute continued, focusing now on what constituted the boundary between Delaware (under the proprietorship of Pennsylvania) and Maryland. Eventually, Maryland and Pennsylvania reached an agreement on Delaware that was tied to an agreement on the Maryland/Pennsylvania border. As a result, Pennsylvania's southern border was relocated 15 miles south of Philadelphia. To assure that the boundaries were surveyed without error, Pennsylvania and Maryland hired two of England's finest scientists, Charles Mason and Jeremiah Dixon.

Pennsylvania's Western Border

With the Mason-Dixon line in place, Pennsylvania could now engage in negotiations with Virginia over its western border. The connection between Virginia and Pennsylvania resides in the fact that, until the Civil War, what is now the state of West Virginia was part of Virginia. And West Virginia does border Pennsylvania, though in what appears to be a creepy way. How did West Virginia come to creep up the side of Pennsylvania? (See Figure 137, in OHIO.)

After the French and Indian War (1754–63), France relinquished its claim to all the land between the Ohio River and the Mississippi River. (See DON'T SKIP THIS.) But England, fearing American expansion that could exceed their control, prohibited the colonists from migrating beyond the Ohio River. The Americans, who had fought side by side with the British, were outraged. The British sought to mollify the Americans and, in one such instance, permitted Virginia to create corporations to invest in land along the Ohio River.

At the same time, Pennsylvania was engaging in the same kind of real estate speculation in the same areas—particularly the region that is today Pittsburgh. (Figure 146) The confusion stemmed from uncertainty

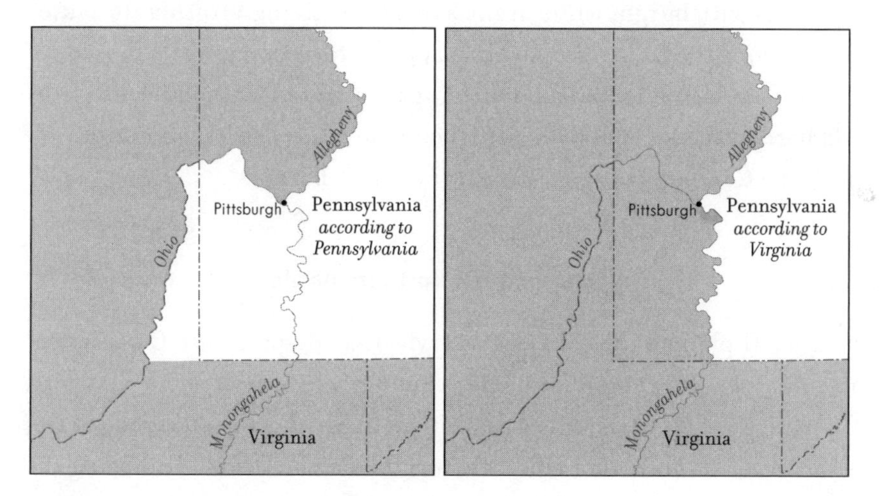

FIG. 146 Pennsylvania/Virginia Border Dispute—1774

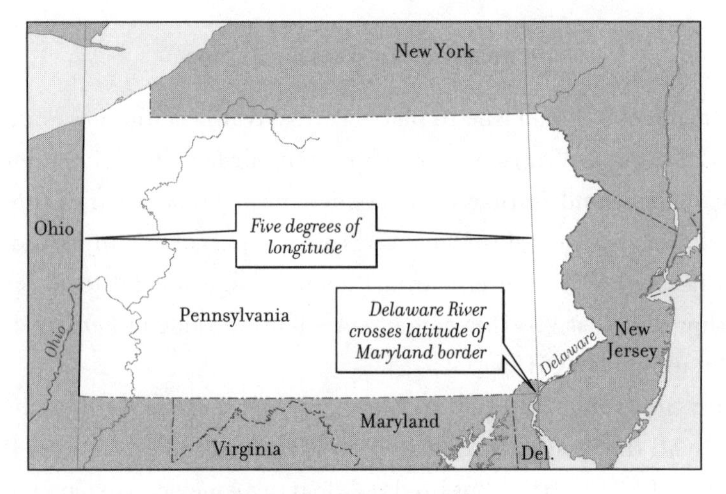

FIG. 147 Pennsylvania Border Agreement

regarding what point on Pennsylvania's eastern border (the Delaware River) is the point from which its western border is five degrees distant.

During the Revolutionary War, Virginia and Pennsylvania arrived at a formula for Pennsylvania's western border. It would be located five degrees west of the point where the Delaware River crosses the latitude of Pennsylvania's border with Maryland. (Figure 147)

This gave Pennsylvania the entryway it sought to the Ohio River, which begins at Pittsburgh, while at the same time giving Virginia its settlements along the Ohio—the part that appears to be creeping up the side of Pennsylvania, but is, in fact, following the Ohio River to the Pennsylvania border. It also enabled Pennsylvania to reach settlements with New York and Connecticut, regarding its northern border.

Pennsylvania's Northern Border

Charles II planted the seed of Pennsylvania's dispute with Connecticut in 1662, nearly twenty years before Pennsylvania existed. In that year, Charles granted a charter to Connecticut in which its western boundary, like that of several other colonies, was the Pacific Ocean. Following the French and Indian War, Connecticut, like Virginia, received permission to form companies to engage in land sales in some of the regions where it had western claims. Connecticut created the Susquehanna Company, which commenced investing in tracts of land in what Connecticut considered to be its newest county and Pennsylvania considered to be the northern third of its colony.

Pennsylvania responded by granting tracts of land in the area to its residents and by asserting its authority to arrest violators of the law—which, by definition, included anyone from Connecticut claiming any of the land. The Connecticut settlers responded, in turn, by building forts for the defense of their settlements. Pennsylvania responded by sending in its militia, to which Connecticut responded with its militia. With all those muskets pointed at each other, it didn't take long before shots were fired and a violent conflict known as the Pennamite War commenced.

The casualties from the Pennamite War would have been considerably greater had not an even bigger war erupted, the American Revolution. The colonies now faced the challenge of acting as a nation. During the Revolution, a special court of arbitration ruled in 1782 against Connecticut, but Connecticut refused to accept the decision. The conflict continued until 1788, when Pennsylvania accepted the ownership titles granted by Connecticut and Connecticut agreed to limit its western land claims to a region along what is today the northern tier of Ohio. (For more on this, go to CONNECTICUT.)

Meanwhile, New York was claiming that its border with Pennsylvania ought to be located along the 42nd parallel, not the 43rd parallel. Pennsylvania's colonial charter stated that its northern border was to be "the beginning of the three and fortieth degree of northern latitude." New York argued that, at the time the charter was written, there were geographers who conceived of latitudes not as lines parallel to the equator but as *bands*, each having one degree of width. The fact that Pennsylvania's charter referred to the "*beginning* of the three and fortieth degree of northern latitude" supported New York claim. If, indeed, the charter did intend for latitudes to mean bands with one degree of width, the beginning of the 43rd parallel would be 42°.

The argument was difficult for Pennsylvania to refute since Pennsylvania itself had used it in claiming that Maryland's charter, locating its northern border at 40°, was in fact 39°—that is, the beginning of the 40th degree (or band) of latitude.

Pennsylvania became more conciliatory in its negotiations with New York once its agreement with Virginia assured Pennsylvania of its possession of Pittsburgh (with its access, via the Ohio River, to the Mississippi River and the sea). In 1785, Pennsylvania released its claim to land north of the 42nd parallel and New York released its claim beyond the westernmost longitude of Lake Ontario. This agreement provided Pennsylvania a window on the Great Lakes that included what is today the port city of Erie—a port that would soon be all the more valuable now that New York, with its newly assured border, could proceed to build the Erie Canal. (See Figure 124, in NEW YORK.)

Seen today, the shape of Pennsylvania appears only mildly interesting. Its straight-line borders make it look more like one of the newer western states than one of the older eastern ones. And, indeed, Pennsylvania's borders do reflect conflicts and their resolutions that took America from being a collection of British colonies to a nation of thirteen united states.

RHODE ISLAND

Rhode Island is such a teensy state, why isn't it just the eastern end of Connecticut? Why does Rhode Island's northern border almost, but not quite, line up with Connecticut's? And why is its straight-line western border located where it is?

When Rhode Island was established, it actually was an island. And the island is still there, out in the middle of Narragansett Bay. It was known then, and is known even now, as Aquidneck Island. (Figure 148)

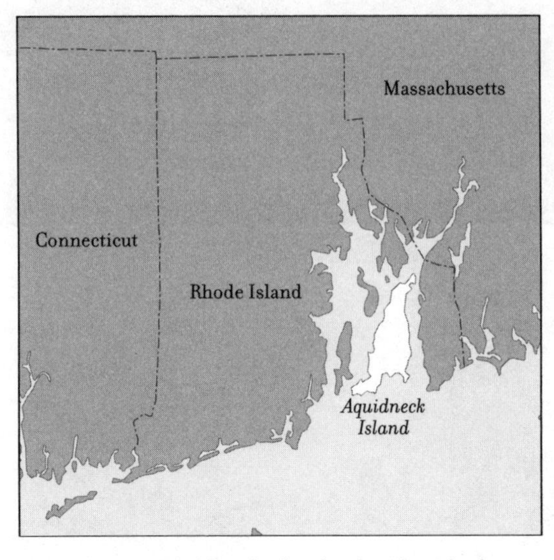

FIG. 148 **Aquidneck Island—the First Rhode Island**

Roger Williams and his followers moved to this island from the Massachusetts Bay Colony to create a new colony based on the then radical notion of religious freedom. Williams and his associates purchased the land from the local Indians. Other land purchases soon followed, and in 1663, King Charles II issued a royal charter for Rhode Island. The boundaries given in the charter caused disputes from the moment it was issued until well into the 19th century.

Rhode Island's Northern Border

According to its charter, Rhode Island's eastern and western borders were to extend to the "south line of the Massachusetts Colony." This description leads to a seemingly simple question: What is the southern border of Massachusetts? Since Massachusetts was an amalgamation of the Massachusetts Bay Colony and the Plymouth Colony, its southern border was subject to interpretation.

Commissioners from Rhode Island and Massachusetts agreed to use the same solution to this problem that had recently been used with Connecticut. The southern border of Massachusetts was deemed to be an east-west line three miles south of the southernmost waterway that feeds into Massachusetts Bay. This was the Neponset River, the waterway that extends farthest south from Massachusetts Bay. And, indeed, its southernmost reach is 3 miles from the northeast corner of Rhode Island. (See Figure 37, in CONNECTICUT.)

Why, then, does Rhode Island's northern border angle slightly to the southwest, and therefore not line up with Connecticut's northern border? Massachusetts sought to preserve its preexisting settlements and deeded land. At the time, it had already granted title to land in what are now the towns of Wrentham, Millville, and Blackstone. Rhode Island agreed to angle its northern border so that it passed below those areas.

The different alignments of Rhode Island's and Connecticut's northern borders preserve a difference in attitudes between the two colonies. Rhode Island, founded by colonists who prized tolerance and understanding, adjusted its border to accommodate its neighbor. Connecticut, founded by the same Puritans who lived in Massachusetts, insisted on their interpretation of the text, resulting in a straighter line—but one with a big scar just west of center.

Rhode Island's Western Border

Under Connecticut's royal charter, its eastern border is Narragansett Bay. Under Rhode Island's charter, issued a year later in 1663, its western border is the Pawcatuck River. Since the Pawcatuck is *west* of the Narragansett, the king foresaw that Connecticut might take offense. So he came up with what he thought was a way to solve the problem. In Rhode Island's charter, Charles II declared

> that the said Pawcatuck river shall be also called Narragansett, and to prevent future disputes that otherwise might arise thereby forever hereafter shall be construed, deemed and taken to be the

> Narragansett River in our late grant to Connecticut Colony mentioned as the easterly bounds of that Colony.

But those keen-eyed Connecticut colonists were not deceived. They continued to contest the land until 1840, when Connecticut finally ratified the boundary. (See Figure 42, in CONNECTICUT.)

Rhode Island's Eastern Border

Rhode Island's remaining boundary conflicts were with Massachusetts, regarding inlets and islands on the eastern side of Narragansett Bay. In 1747, King George II issued a decree determining which of the contested areas belonged to Rhode Island and which to Massachusetts. (Figure 149)

Some conflicts continued, however. Rhode Island maintained it should have jurisdiction over Pawtucket and East Providence. Massachusetts maintained that Fall River had traditionally belonged to it. In 1862, the two states agreed to exchange these regions. (Figure 150)

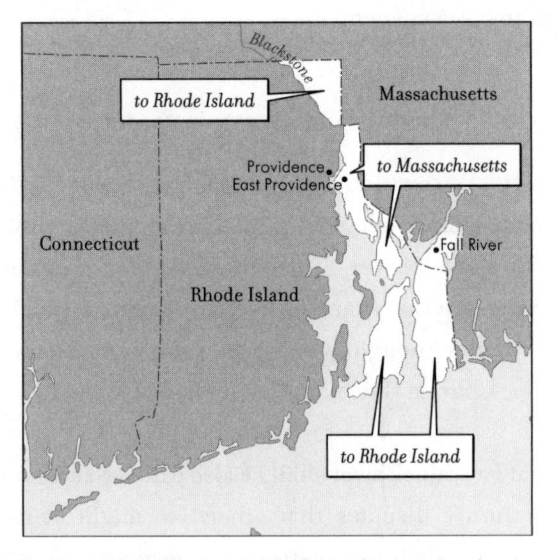

FIG. 149 Rhode Island/Massachusetts Border Ruling—1747

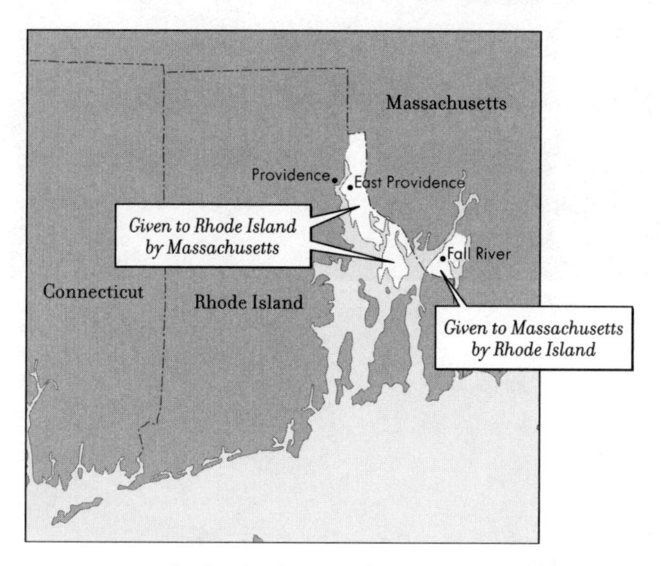

FIG. 150 Rhode Island/Massachusetts Land Swap—1862

Rhode Island is the only state in the union founded expressly for religious freedom. It was the first American colony to celebrate, rather than tolerate, the differences among us. In this respect, the fact that Rhode Island is also the smallest state in the union gives one pause.

SOUTH CAROLINA

Why is South Carolina separate from North Carolina? And since they are separate, why is South Carolina so much smaller? How did South Carolina's border with North Carolina end up having so many angles and steps? There's even part of a square at one point—where did that come from?

South Carolina was originally joined with North Carolina as, simply, he Carolina Colony. King Charles I issued Carolina's initial charter in

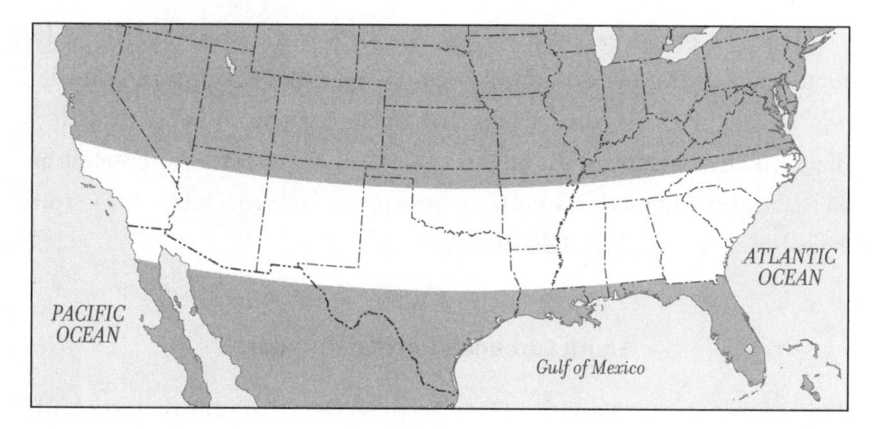

FIG. 151 The Carolina Colony—1629

1629 to reward a political ally named Robert Heath (often erroneously listed as Richard Heath, a British judge of that era). The king granted to Heath all the land between the St. Mathias River (now known as the St. Marys River) on the south, the middle of Albemarle Sound on the north, the Atlantic Ocean on the east, and the Pacific Ocean on the west. (Figure 151)

Heath never developed his property, and in time the grant became void. In 1663, when King Charles II granted Carolina to a group of his political allies, he expanded the southern border his father had established, locating it at the 29th parallel, which crosses Florida some 20 miles south of what is today Daytona Beach. This adjustment reflected the fluidity of England's territorial contest with Florida's claimant, Spain.

Two years later, Charles II relocated Carolina's northern border to settle a dispute between Carolina and Virginia. The king moved the border one-half of one degree farther north, giving all of Albemarle Sound to the Carolina Colony. The boundary was now midway between Albemarle Sound and the Chesapeake Bay. (For more on this dispute, go to NORTH CAROLINA.) Carolina's new northern border, 36°30', would go on to a long and, with Missouri's statehood, important life in American history. (See DON'T SKIP THIS.) But it began here, as a minor colonial adjustment.

Unlike its predecessor, this Carolina Colony did attract settlers. Its population clustered around two regions: Albemarle Sound and Charleston harbor. The distance separating these regions, along with differences in the background and prosperity of their settlers, created an increasing strain on the colonial government. In 1710, Queen Anne split the colony into North and South Carolina.

South Carolina's Northern Border

Queen Anne sought a dividing line between the Carolinas that was halfway between the colony's northern and southern border. England viewed the southern border, at that time, as being the Savannah River—well to the north of either previously stipulated border. To this day, the Savannah River serves as the southern border of South Carolina. Halfway between the Savannah River and 36°30' is the Cape Fear River. (See Figure 126, in NORTH CAROLINA.)

The Cape Fear River would have been a fine boundary between the two colonies, except that North Carolina, which had previously been granted a degree of self-rule, had already issued titles to land on both sides of the river. The problem remained unresolved until 1730, when the two Carolinas agreed to a new border. Since South Carolina was more prosperous at the time than North Carolina, it released its claims to the Cape Fear River. The new line began 30 miles down the coast from the Cape Fear River and headed northwest until it reached 35° N latitude. At this point, the line was to go due west to the Pacific Ocean. (See Figure 127, in NORTH CAROLINA.)

This line, however, ran into trouble. Or, more specifically, it ran into Indians—the Catawba Indians—who at the time were allied with the colonists and England against French encroachment from the Mississippi River regions. So in 1735, the two sides tried again. The line remained the 35th parallel, but this time it was to make an adjustment if it arrived at the Catawba lands. Amazingly, the line did not interfere with the Catawba lands since, as they later learned, the surveyors had mistakenly located it 13 miles to the south. (See Figure 128, in NORTH CAROLINA.)

The series of steps and notches in South Carolina's northern border can be traced to this erroneous boundary. It was left in place when, in 1771, the border was continued farther west. To compensate South Carolina for its loss of land, England ruled that the extended boundary would turn north at the Catawba land, follow its perimeter to the Catawba River, follow that river to its first branch, then head due west. (See Figure 129, in NORTH CAROLINA.) The line due west compensated South Carolina not only because it was located north of the 35th parallel but also because this time the surveyors mistakenly veered even farther northward as they located it. To this day, South Carolina's northern border contains these errors and adjustments.

Still, a big question remains: Why did South Carolina accept any of these proposals, since they all divided the colony in a way that left South Carolina so much smaller than North Carolina? It did because, at the outset of these border adjustments in 1710, the division of the colony did not make South Carolina smaller than North Carolina. (Figure 152) A later adjustment to South Carolina's southern border did that.

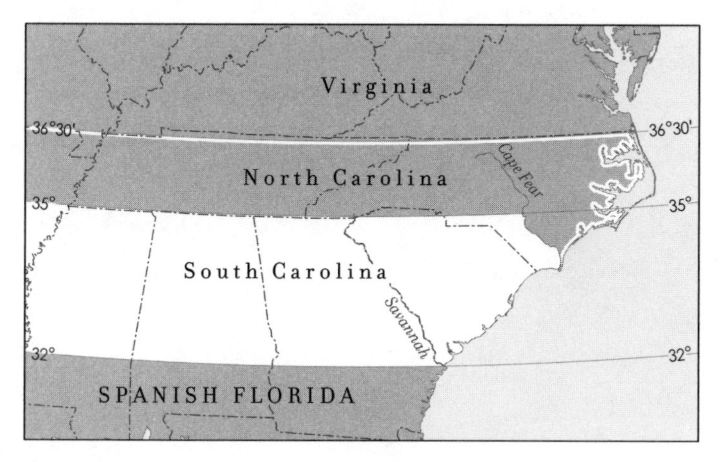

FIG. 152 Size of North and South Carolina (in Theory)—1710

South Carolina's Southern Border

In 1732, King George II created the colony of Georgia. The boundaries specified in the charter not only gave Georgia all of its land from South Carolina but also cut off South Carolina's access to land farther west by declaring the Savannah River to be their border. It was the creation of Georgia that resulted in South Carolina's becoming a smaller colony (and later state) than North Carolina. But the creation of Georgia also established a buffer between South Carolina's wealthy rice and indigo plantations and the threat posed to them by the Spanish in neighboring Florida. For this reason, South Carolina invited the founder of Georgia, James Oglethorpe, to address its legislature and welcomed the new colony as its neighbor to the south.

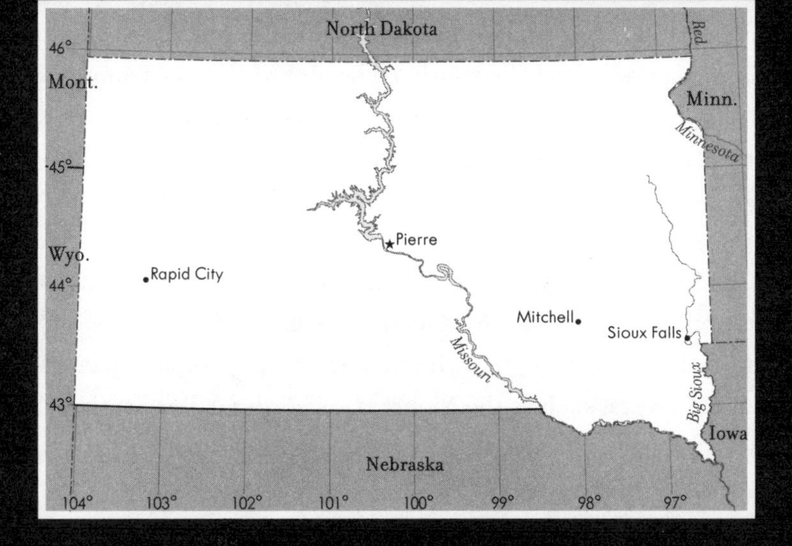

Why is the straight-line section of South Dakota's eastern border not lined up with the northwest corner of Iowa? Why is South Dakota's southern border a straight line when the Niobrara River runs parallel to it, just below? And why are South Dakota's straight-line borders located where they are?

After Minnesota became a state, Congress combined the land left over from the Minnesota Territory with other territorial land and created the

Dakota Territory. (See Figure 133, in NORTH DAKOTA.) In the process, what would become the eastern and southern borders of South Dakota first surfaced. But South Dakota's southern border had, in effect, been established seven years before when Congress created Kansas, a state that doesn't even border South Dakota.

South Dakota's Southern and Northern Borders

In the course of the controversy surrounding the 1854 Kansas-Nebraska Act, Congress adjusted the southern border of Kansas from 36°30' to 37°. In doing so, Congress made it possible for a tier of four states, each having three degrees of height, to fit between Oklahoma and Canada. Those states became Kansas, Nebraska, South Dakota, and North Dakota.

By this math, South Dakota's southern border should be located at 43° and its northern border at 46°. And indeed the southern border of South Dakota is 43° (revealing that the mathematical plan superseded even convenient rivers, such as the Niobrara, as borders). But South Dakota's northern border is located at 45°55'—one-twelfth of one degree short. Why the slight discrepancy?

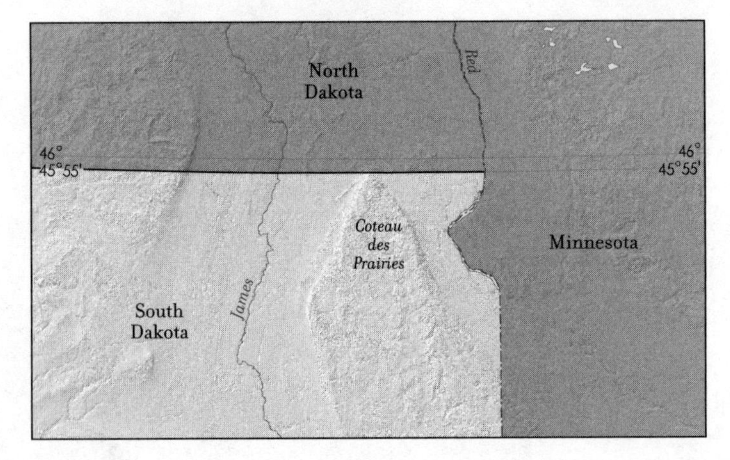

FIG. 153 The North and South Dakota Border—the Reason for Deviation

The two states agreed to locate their boundary at the northern tip of Coteau des Prairies, a 200-mile-long plateau. (Figure 153) This location had the advantage that the boundary's location would now be obvious. Also, the adjustment helped to equalize the size of the two states, since North Dakota's eastern border, the Red River of the North, veers slightly westward, narrowing that state's overall size.

South Dakota's Eastern and Western Borders

South Dakota inherited its eastern border from the state of Minnesota. This border combines a straight line, due north, with a series of waterways that, taken together, traverse an essentially north-south line. (For more on this border, go to MINNESOTA.)

South Dakota also inherited its western boundary, in this instance from Idaho—once again, a state that does not border either of the Dakotas. After the creation of the state of Washington in 1889, Congress reorganized the territories. It combined the land left over from the Washington Territory with land from the western reaches of the Dakota and Nebraska territories and created the Idaho Territory. As it turned out, the boundaries of the Idaho Territory were unworkable, and it was soon divided yet again with the creation of Montana. But the eastern border of this short-lived Idaho Territory, located along the 104th meridian, established what would become the western border of South Dakota. (See Figure 134, in NORTH DAKOTA.)

Why did Congress locate this border where it did? Why not the 105th meridian? The reason was very much like that which determined the southern border of South Dakota. By locating the border at 104°, three future states could be created along this meridian, each having almost exactly seven degrees of width. Those states turned out to be South Dakota, North Dakota, and Wyoming. Moreover, they joined three other western states—Colorado, Washington, and Oregon—each of which also has almost exactly seven degrees of width. (See Figure 13, in DON'T SKIP THIS.)

South Dakota and North Dakota could well serve as Exhibit A in demonstrating the longstanding effort by Congress to create states that are equal. In fact, so powerful was this concept, that President Benjamin Harrison shuffled the documents creating North and South Dakota in 1889 before signing them so that no one would know which of them came into existence first.

Why is there a notch in the northwest corner of Tennessee? Why are its northern and southern straight-line borders located where they are? And why is its northern straight-line border not exactly straight?

Tennessee's Northern Border

The land that is today Tennessee was previously part of the Carolina Colony. It was during this time that the basis of what would become the northern border of Tennessee first surfaced. In 1655, King Charles II reset the border between Virginia and Carolina at 36°30'. (To find out why the king chose that particular latitude, go to NORTH CAROLINA.) To this day, 36°30' remains the northern border of Tennessee. Sort of. When the state of Tennessee was created, this issue would be revisited.

Tennessee's Southern Border

When the Carolina Colony was divided into North and South Carolina in 1712, Tennessee's present-day southern border emerged. Even though the division of the Carolinas turned out to be a messy affair, through it all both sides continued to aim for the 35th parallel. Though they never did get it right, the 35th parallel was still considered the division of their lands beyond the Appalachians. Thus, to this day, the 35th parallel serves as the southern border of Tennessee.

Tennessee's Eastern Border

Following the Revolution, North Carolina joined other states with extensive colonial claims in donating those lands to the United States government to create more equal sized states (and to create more pro-slavery or anti-slavery states to maintain each side's maximum voting power in the Senate). In ceding the land that would become Tennessee, North Carolina needed to define the boundary between it and the state it was spawning. The Appalachians provided the obvious border, but where, amid their many peaks and valleys, was the best location for the line? North Carolina opted for the highest crests, which serve to this day as the eastern border of Tennessee.

Except at its southeast corner. There, for mysterious reasons, the border departs from the crests and heads straight down to Georgia. (To find out more, go to NORTH CAROLINA.)

Tennessee's Western Border

When what is today the state of Tennessee was still part of England's Carolina Colony, the western border of that colony was, on paper, the Pacific Ocean. On earth, however, Indians, Frenchmen, and Spaniards occupied the land between the Carolina Colony and the beach at Big Sur. After the Revolution, when the new United States needed to define its borders, it declared the Mississippi River to be its western border. Thus, when North Carolina ceded the land that became Tennessee, the western border of Tennessee was the Mississippi River. However, at its western end, the land from the Tennessee River to the Mississippi River belonged, by treaty, to the Chickasaw Indians.

Tennessee's Northern Border Revisited

Tennessee's northern border—supposedly the 35th parallel—contains several deviations and one major jog to the south at its western end. Why? (See Figure 77, in KENTUCKY.)

In creating Tennessee and Kentucky, their parent states, North Carolina and Virginia, commissioned surveyors to extend their mutual border to the west. The task, however, proved to be more difficult than expected. Some of the deviations in Tennessee's northern border pertain to the fact that the line it inherited, the Virginia/North Carolina boundary, turned out to be less than accurate. Though it was not known at the time of Tennessee's creation, the Virginia/North Carolina border had been located slightly north of 36°30' all the way back at its starting point on the Atlantic coast. In addition, it veered slightly northward as it extended to the west.

Other deviations in Tennessee's northern border preserve disagreements between the Virginia and North Carolina surveying teams. These disputes began almost immediately when North Carolina's surveyor, Colonel Richard Henderson, claimed that iron deposits in the mountains were deflecting the compass of Virginia's surveyor, Dr. Thomas Walker. Walker disagreed and the two teams separated. Left to his own devices,

Walker generated a boundary line that curved slightly southward, then northward.

In 1802, Tennessee and Kentucky agreed to a Compromise Line. But in seeking to implement the Compromise Line, new disputes surfaced, many having to do with uncertainties as to the location of Walker's line, whose markers were beginning to disappear. As had been the practice in his day, Walker had used mostly trees that he'd marked to designate the boundary. Whether through natural causes or not, many of the trees were no longer standing.

In 1856, Tennessee and Kentucky agreed to try again to define the boundary between them. "We began the experimental work at the town of Bristol," the surveying team wrote to Tennessee's governor, "a small village situated on the Compromise Line of 1802, at a point where there was no controversy as to the locality of the line." Years later, Bristol would be the cause of yet another change in Tennessee's northern border. For now, however, it was simply a convenient marker for the surveyors. But the 1856 survey failed to end the controversy.

The ongoing dispute contributed to the most prominent deviation in Tennessee's northern border, the sudden drop in the line just west of the Tennessee River. This location is where Walker's line ended, since the land beyond belonged at the time to the Chickasaw Indians. In 1819, General Andrew Jackson, on behalf of the United States government, purchased from the Chickasaw their land east of the Mississippi River, south and west of the Tennessee River, and north of the Mississippi state line. (Figure 154) This region, known as the Jackson Purchase, was to be annexed to the states of Tennessee and Kentucky. Because it was divided along the *actual* 36°30', it drops below the line leading to it, which had veered to the north.

In 1901, the United States Supreme Court adjudicated the remaining disputes over the Tennessee/Kentucky border. As a result, the eastern end of Tennessee's northern border now begins at 36°36', proceeds west for 15 miles, then drops to the 1802 Compromise Line. (Figure 155) Why did the Supreme Court sanction this?

The Court sanctioned it because Tennessee agreed to it. Overall,

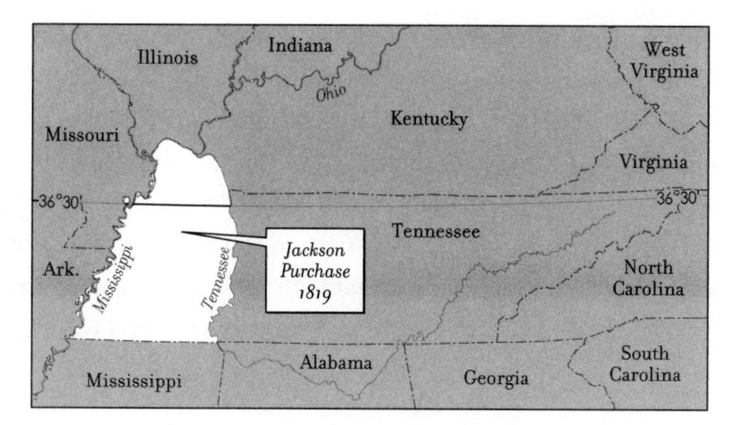

FIG. 154 The Jackson Purchase—1819

Tennessee came out the victor in the case. Walker's erroneously located boundary was generally 10 miles north of where it should have been. The formula for the 1802 Compromise Line improved the situation for Kentucky but did not undo the error. But one result of the error Tennessee did undo: it ceded its half of Bristol, the little town that straddled the Compromise Line. As a result, Tennessee created the jog in its border at the eastern end. Prior to 1901, the Compromise Line was a straight line from the Virginia/North Carolina border to the Cumberland Gap. After 1901, the line began as it had, but after passing through the Appalachian crests, it now turned southward for 2 miles then went westward to the Cumberland Gap, so that it passed below the town of Bristol.

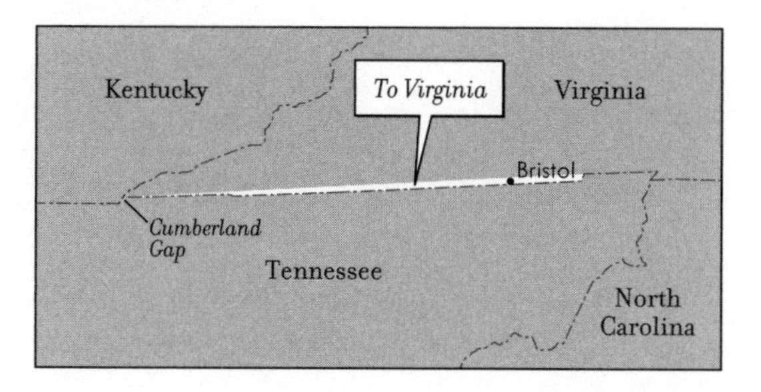

FIG. 155 The Jog in Eastern Tennessee from Land Cession to Virginia—1901

Other irregularities farther along Tennessee's northern border resulted from properties whose deeds were recorded in one state or another, based on which side of which line the buyers and sellers believed themselves to be, or from personal reasons—some individuals preferred to be residents of one particular state or county. Local lore asserts that when the Walker line was redrawn, first in the 1820s and then in the 1890s, it was not unheard of for the residents to bribe the surveyors, frequently with locally made moonshine. This may well explain why Tennessee's northern border staggers.

Why does Texas have that northern panhandle? And since it does have that panhandle, how come it stops short of Kansas, creating an even smaller panhandle in Oklahoma? Why does the western border of Texas stop following the Rio Grande where it does, and why does it

Texas' Eastern Border

Texas first became a place of continual colonial settlement in 1691, when Spain grew alarmed at reports that Frenchmen had crossed the Sabine River from Louisiana. The French, possibly testing the waters for colonial expansion, were befriending the local Indians, an alliance of tribes known by their native word for "allies," *tejas*. Spain dispatched an expedition to clear the area of the French and convert the Indians to Christianity. To ensure that the French stayed out and the Indians stayed Christian, Spain built missions throughout the region and established the province of Tejas. Its eastern border at the time was the Sabine River, which to this day is a segment of Texas' eastern border.

Spanish worries regarding French incursions into their North American domains came to an end in 1803, when the United States purchased the Louisiana Territory. Now Spanish worries regarded the Americans. In 1819, the United States and Spain concluded the Adams-Onis Treaty, defining the boundary between American and Spanish territory. (See DON'T SKIP THIS.) A good deal of today's Texas border was established in this treaty. And a number of the border's segments also preserve elements in the history of Spanish Tejas.

The current border preserves not only the original eastern border of Tejas along the Sabine River but also the border to which the province later advanced, the Red River. (Figure 156)

But why is this advance represented at the point where the Sabine crosses the 32nd parallel? Why not some other degree of latitude?

In the negotiations between Secretary of State John Quincy Adams and Spanish envoy Lord Don Luis de Onis, the Red River was a particularly contentious element. It had become a boundary of Tejas when Spain advanced northward from the Sabine River to the Red River. For the United States, the Red River was an important avenue of commerce in Louisiana, flowing diagonally from its northwest corner across the territory to the Mississippi River.

By locating the border between the Sabine and the Red rivers by a vertical line heading north from the point at which the Sabine River crosses

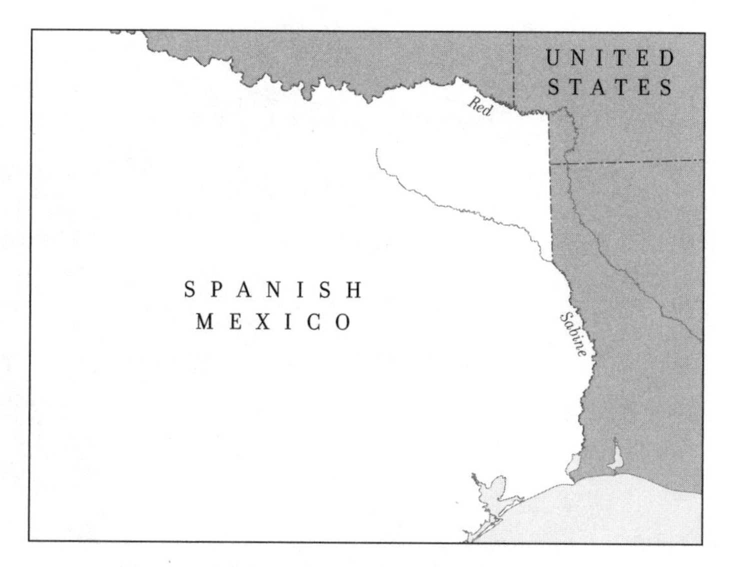

FIG. 156 Spanish Expansion Preserved in Texas' Border

the 32nd parallel, the border that results gives the United States the entire diagonal segment of the Red River, along with a buffer west of the river which, at its narrowest point (in present-day Arkansas) is exactly 10 miles wide. In addition, the point at which this vertical line intersects the Red River gives Spain access to the rest of the Red River along its southern bank. (See Figure 80, in LOUISIANA.)

We can also see in today's eastern border of Texas another segment from the Adams-Onis Treaty, which specified that the U.S./Spanish boundary continued westward along the Red River to 100° W longitude, at which point it turned due north to the Arkansas River. Today, what remains of this line due north is the eastern edge of the Texas panhandle. (Figure 157)

But why did they pick the 100th meridian as the point at which the border would depart from the Red River and head north? Part of the reason is revealed in a letter Onis had written to Adams in February 1819:

> I have to state to you that his majesty is unable to agree to [the boundary of] the Red River to its source, as proposed by you. This

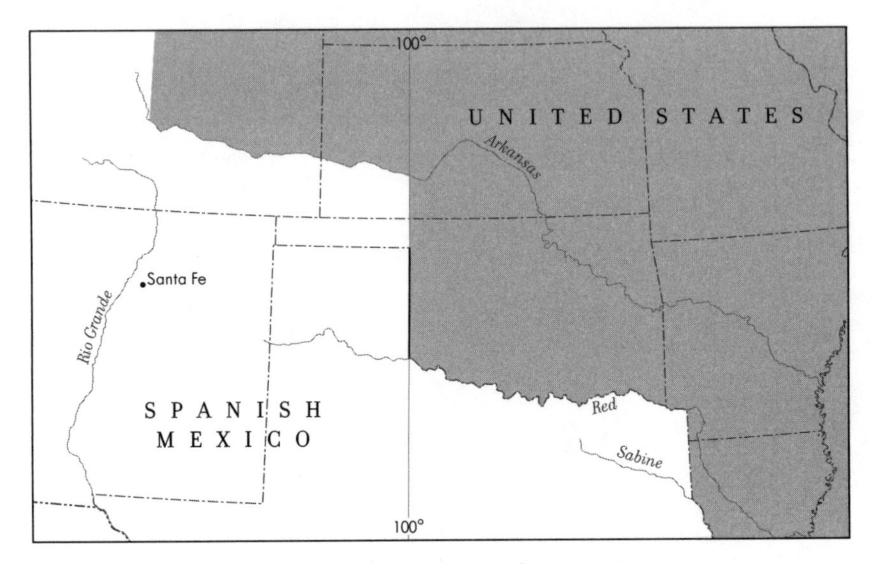

FIG. 157 The Texas Panhandle—Spain Protecting Santa Fe

> river rises within a few leagues of Santa Fe, the capital of [Spanish] New Mexico.

Ultimately, the two sides agreed upon a border farther north at the Arkansas River. For the point at which the boundary would jump from the Red River to the Arkansas River, they selected the midpoint between the headwaters of the Arkansas River and the point where it crosses into what was already the state of Louisiana. That midpoint is 100°. (Figure 158)

Oklahoma would later dispute the Red River segment of the Texas boundary, claiming that the north branch of the Red River, being the smaller branch, was not the legitimate boundary of Texas. In 1897, the Supreme Court agreed, shifting the Texas boundary to the south branch of the Red River. (See Figure 139, in OKLAHOMA.)

Texas' Northern Border

Spain might have saved itself the effort of negotiating the Adams-Onis Treaty had it known that, only two years later, Mexico would win its inde-

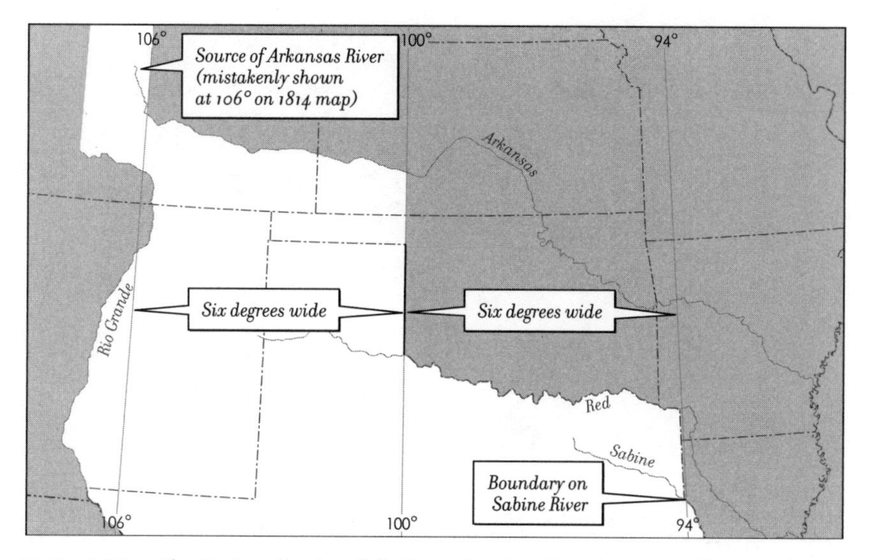

FIG. 158 The Eastern Border of the Texas Panhandle—the Reason for Location

pendence. The worry of American incursion now belonged to the Mexicans, who continued the Spanish policy of prohibiting American settlements within their territory. But Moses Austin was able to get the Mexican government to grant him an exception, allowing him to establish a lead-smelting operation in the Mexican province of Texas. Austin's son, Stephen, took over the business, the success of which led Mexican authorities to permit other American ventures. By 1830, Americans outnumbered Mexicans in the province, and in 1836, Texas achieved independence under the leadership of Sam Houston.

The northern border of the Republic of Texas was strangely shaped but quite logical. (Figure 159) It followed the boundaries set in the Adams-Onis Treaty. As things turned out, being its own country was a lot more expensive than Texas had anticipated. Deeply in debt, Texas joined the United States in 1846.

Since, in the Republic of Texas, slavery was legal, the U.S. state of Texas had to comply with the Missouri Compromise if it wished to remain a slave state. This 1820 agreement prohibited slavery north of 36°30'. Texas therefore relinquished to the United States all of its land

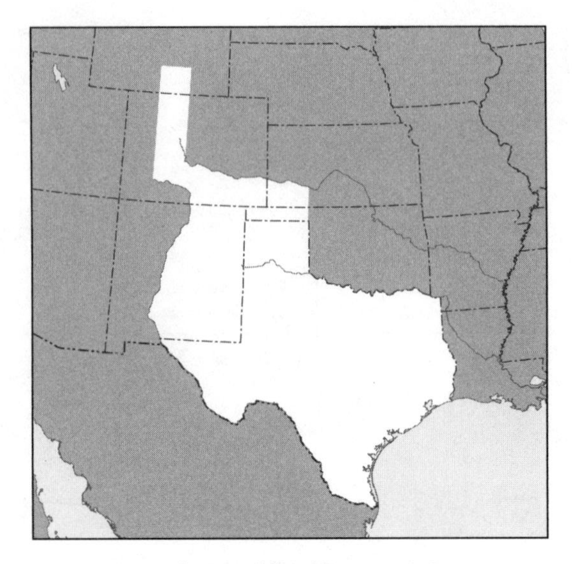

FIG. 159 The Republic of Texas—1836

north of that latitude. This is why what would become the Texas panhandle is sliced off where it is, forming the northernmost border of Texas. (See Figure 73, in KANSAS.)

Texas' Southern Border

The 1836 treaty that ended the Texas War of Independence stipulated that the Mexican army would withdraw south of the Rio Grande. The Rio Grande had never been the southern border of Texas in its days as a province of Mexico, but it was employed in the treaty to provide security. Texas anticipated that the river would become its officially recognized border. But Mexico never recognized the Republic of Texas. When Texas joined the Union in 1846, the United States stipulated that the southern border of the state of Texas would be the Rio Grande. Mexico, fearing that Texas might be the preamble to further American expansion, reasserted its claim to the entire region and backed that claim with its army. The United States did likewise, resulting in the Mexican War. In the 1848 treaty that followed Mexico's surrender, the Rio Grande became the officially recognized southern border of Texas.

Texas' Western Border

Texas continued to claim all its land south of 36°30', resulting in a state that so dwarfed every other state it undermined the principle that all states should be created equal. Congress passed a law stating that, if it desired, Texas could subdivide itself into as many five states. Southerners in particular were attracted to this idea, desiring the additional slave states to help represent their cause in the U.S. Senate. But Texas wasn't wild about the idea.

Texas, however, remained burdened by crushing debts from its days as a republic. For this reason, under the Compromise of 1850, Texas sold the United States all of its land west of 103° and north of what was then the latitude of New Mexico's southern border. (See Figure 119, in NEW MEXICO.) This sale resulted in the right-angled western border of Texas that we see today. (To find out why Congress wanted the land starting at the 103rd meridian, go to NEW MEXICO.)

Even with the land Texas surrendered north of 36°30' and the land it later sold, Texas remains far larger than most other states. If Congress had been truly committed to the notion that all states should be created equal, how had this happened?

Congress did not create Texas. Texas created Texas. And when the opportunity presented itself to bring Texas into the United States, it was at a time when many Americans believed it to be the nation's destiny to possess all of North America. The opportunity to acquire Texas was too great to risk by imposing conditions regarding its division into smaller states. Congress did, however, require—and later persuade—Texas to release large sections of its land. These actions affirm, rather than refute, the notion that Congress sought to create states that were equal.

UTAH

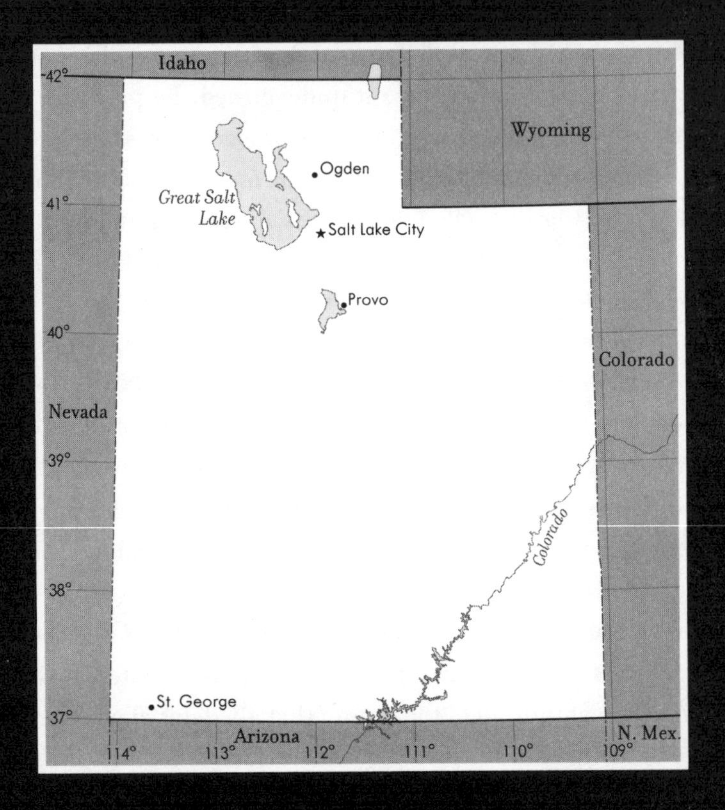

Why isn't Utah a rectangle? How come Wyoming got to bite off a corner of Utah instead of the other way around? Aren't the rest of Utah's borders located where they are simply because they line up with those of neighboring states?

Utah's Northern Border

The northern border of Utah first began to surface before the land that is now Utah was even part of the United States. The segment of Utah's northern border located along the 42nd parallel is an extension of Spain's coastal boundary with England from their 1790 Nootka Convention. It was later extended in 1819, when Spain and the United States concluded the Adams-Onis Treaty, stipulating the borders between Spanish territory and American territory acquired in the 1803 Louisiana Purchase. (For more on both of these agreements, see DON'T SKIP THIS.) Thus, when the region that includes Utah came into American possession in 1848, following the Mexican War, the future Utah inherited the 42nd parallel as its northern border—with the exception of its northeast corner, where Wyoming's rectangle juts across the line.

Just as the Mexican War was ending, the Mormons were packing their bags in Nauvoo, Illinois, where their practice, at that time, of polygamy had resulted in conflict with the authorities. The Mormons headed west, crossing the Rocky Mountains in search of land where they could live beyond the prying of hostile authorities. They settled at the Great Salt Lake, believing they had found such a land. But the boundaries of Utah are evidence of the futility of trying to hide from Uncle Sam.

Utah's Southern Border

Almost immediately after settling at the Great Salt Lake, Mormon leader Brigham Young issued a call to all the citizens east of the Sierra Nevada and west of the Continental Divide to meet in Salt Lake City to discuss the political needs of this vast, largely arid, realm. The result of that convention (attended mainly by his fellow Mormons) was a proposed state, to be called Deseret. (Figure 160)

Congress, however, had ideas of its own, and they didn't include creating a state far larger than California, the size of which was itself controversial. In 1850, Congress created the Utah Territory. The new territory was nearly as wide as the proposed state of Deseret, but its height was limited on the south by the 37th parallel and on the north by the preexisting 42nd parallel. (Figure 161)

The 37th parallel is the same southern border as that of Utah's eastern

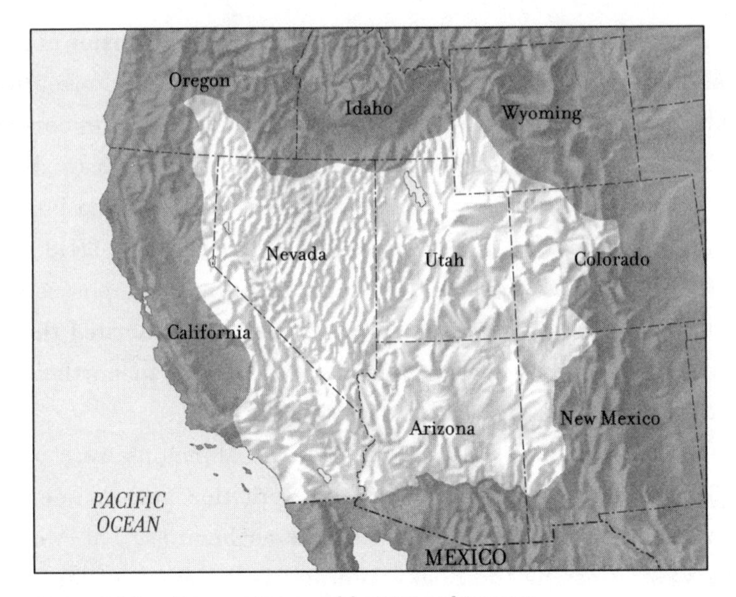

FIG. 160 Mormon Proposal for State of Deseret

neighbor, Colorado. Congress may have envisioned a tier of states, each having four degrees of height, just as Colorado, Wyoming, and Montana came to be. (Figure 162) But three years later, Congress precluded this possibility by dividing the Oregon Territory horizontally, because of objections by those citizens to a vertical border along the crest of the Cascade Mountains.

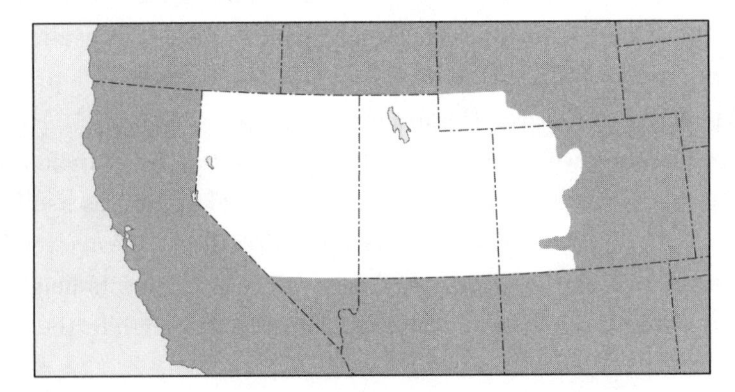

FIG. 161 The Utah Territory—1850

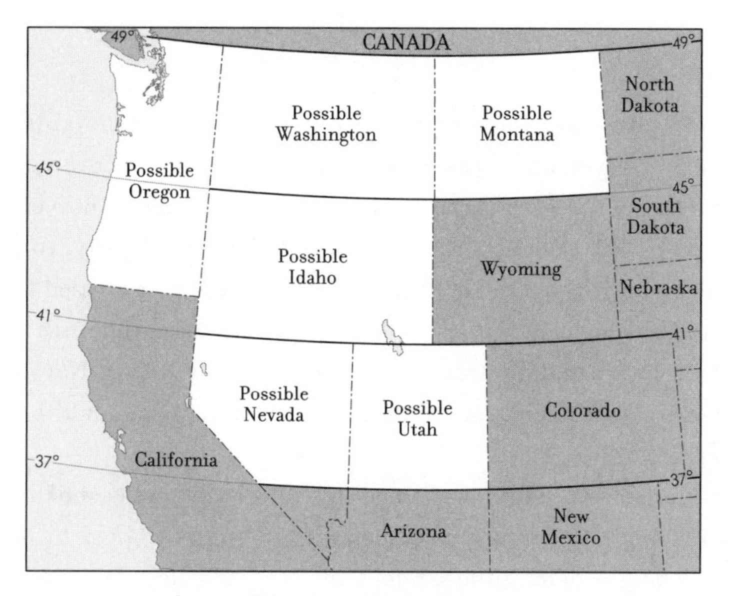

FIG. 162 **The Possibility for a Second Column of States with 4° of Height**

Utah's Eastern Border

In 1858, gold was discovered in the Rocky Mountains in a region that was then the western end of the Kansas Territory. Within a year, residents freshly arrived in the area proposed the formation of the "Territory of Jefferson." As with the proposed "Territory of Deseret," Congress responded by accepting the idea while changing both the name and the boundaries. Where Deseret was renamed Utah, Jefferson became Colorado. And like Utah, Colorado's boundaries were adjusted to very particular locations. (For details, go to COLORADO.) The western border of Colorado took a hefty slice from the Utah Territory. Utah's eastern border had been the crest of the Rockies. With the creation of Colorado, it became the 109th meridian. This meridian remains Utah's eastern border to this day.

By locating the eastern border of Utah at 109°, Colorado ended up with seven degrees of width, as have Wyoming, North Dakota, South Dakota, Washington, and Oregon. And Utah—but only briefly.

Utah's Western Border

At the same time that gold was discovered in the Rocky Mountains, a vast discovery of silver, known as the Comstock Lode, was discovered in the western end of the Utah Territory. In 1861, with the nation now engaged in the Civil War, Congress was uncertain of Mormon loyalty. To assure possession of these lucrative mining regions, Congress created the territory of Nevada. Its boundary with Utah was located at the 116th meridian. This location both secured the silver mines within Nevada and resulted in Utah having seven degrees of width. What could be more perfect?

But more silver, and now also gold, were discovered east of 116° the following year. Under the circumstances, Congress opted to deprive Utah of a degree of longitude, relocating the Utah/Nevada border at the 115th meridian. (See Figure 110, in NEVADA.)

In 1864, Nevada was granted statehood. At Nevada's request, Congress sliced another degree of longitude off Utah, to give Nevada, which is predominantly desert, access to waterways leading to the Colorado River. The 114th meridian has since remained the western border of Utah.

Utah's Northern Border Revisited

Yet another bite would be taken out of Utah, this time by Wyoming when it became a territory in 1868. Why did Congress create a border that deprived Utah, or rewarded Wyoming, of land that had been Utah's northeast corner?

Three factors may have contributed to Congress opting to let Wyoming have the squared-off corner. First, it gave Wyoming seven degrees of width. Second, for those in Congress who remained concerned about Mormon influence and affluence, this corner would have raised concern, since it was particularly rich in resources. There were not only coal fields but also numerous waterways. And there were roadways—the Oregon Trail, the California Trail, and the Overland Trail all passed through this region along with the transcontinental telegraph, the stagecoach

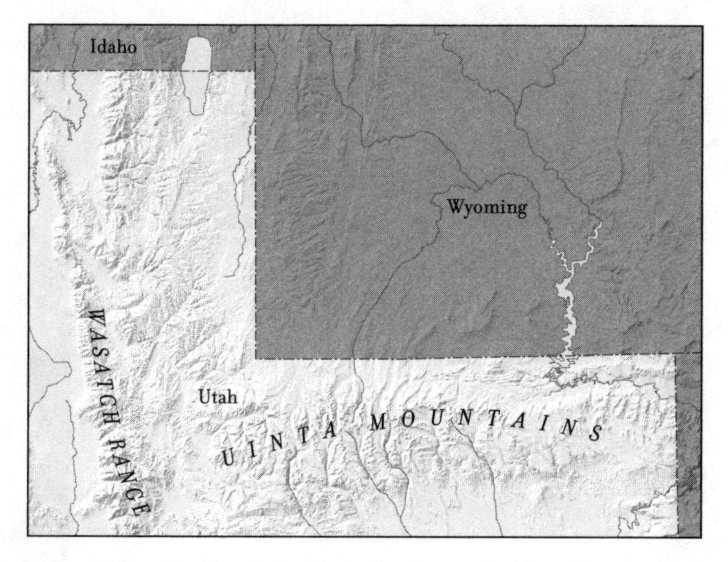

FIG. 163 The Mountains Enclosing Former Northeast Corner of Utah

express, and, starting the same year that Wyoming was created, the transcontinental railroad.

Even for members of Congress no longer so concerned about the Mormons, there remained a geographic factor favoring the annexation of this corner to Wyoming. It is virtually walled off from Utah by the Uinta Mountains. (Figure 163) Had it remained the northeast corner of Utah, access to this pocket would have been difficult for state authorities, whereas access for Wyoming was wide open.

Utah is the only state that Congress created with boundary adjustments that made it *less* equal than others. This blemish in our state borders preserves the fact that blemishes on our nation's ideals are indeed a part of our history. And yet, in the context of the entire American map, Utah's borders also reflect the fact that such blemishes are the exception, not the rule.

VERMONT

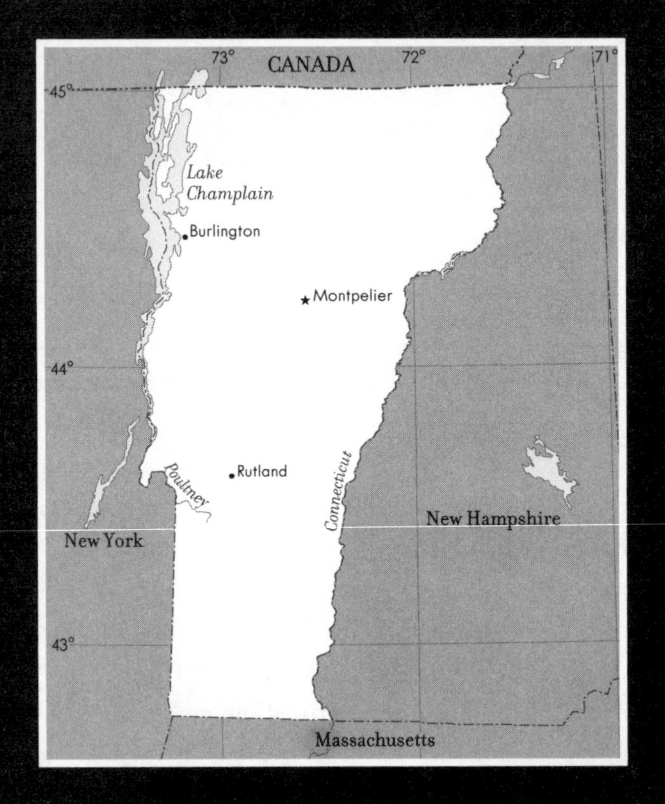

How come Vermont has to stop at a straight line when New Hampshire gets to keep on going north? Since Vermont and New Hampshire together form a nice rectangular shape, why didn't they form a single, more normal-size state?

The reason Vermont is not New Hampshire's western half is rooted in overlapping land claims between the Dutch and the British. In 1616, the Dutch laid claim to all the land between the Connecticut and Delaware

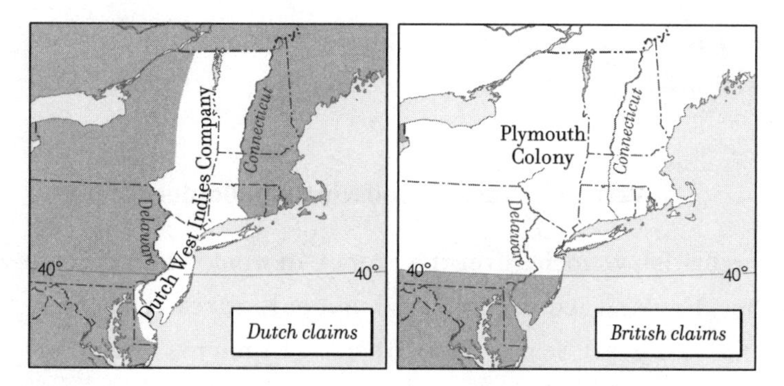

FIG. 164 **Overlapping Claims to What Is Now Vermont**

rivers. What is now known as Vermont was included in these boundaries. Four years later, King James I of England issued a charter to the Plymouth Company laying claim to all the land from 40° to 48° N latitude between the Atlantic Ocean and the Pacific Ocean. (Figure 164)

In 1664, England drove the Dutch out of North America, and what had been the New Netherlands (see Figure 114, in NEW JERSEY) became the colony of New York (which, at the time, included proprietorship over what is now New Jersey) and the colony of Pennsylvania (which, at the time, included proprietorship over what is now Delaware). But the boundary disputes did not depart with the Dutch, since the colony of New York assumed its boundaries were those of its Dutch predecessor. Specifically, New York maintained that its eastern border was the Connecticut River. Its neighbors to the east disagreed.

Vermont's Southern Border

The first of Vermont's borders to surface was its southern border, back when what is now Vermont was part of New Hampshire—or part of Massachusetts, depending on which side you take in the New Hampshire/Massachusetts colonial dispute. King George II took New Hampshire's side and, in 1741, ruled that the southern border of New Hampshire was the line that followed the Merrimack River from a distance of 3 miles to its north, continuing to the point where the river turns north. From here a due west line takes over, ending at the eastern border of New York. This due west line later became the southern border of Vermont.

Vermont's Eastern and Northern Borders

After the British victories over the French in what is today the province of Quebec, England acquired, in 1763, not only a great deal of new territory but also a great deal of new subjects. Unhappy, French-speaking subjects. Without France as their protector, the Quebecois worried about, among other things, the expansionist (not to mention Protestant) Americans to their immediate south. England, for its part, worried about any conflict that might cause disruption of commerce along the St. Lawrence River, for which England had fought so hard. Securing a separation between the American colonies and Quebec seemed like a good idea.

In 1763, King George III declared that the 45th parallel, from the St. Lawrence River to the Connecticut River, would to be the border between the province of Quebec and the colony of New York. More important, the king declared the Connecticut River to be the eastern border of New York—not, as New Hampshire claimed, the Hudson River. (Figure 165)

Why would the king do this? Certainly New Hampshire needed the territory more than New York. Not only was New York larger than most other colonies but, with the Hudson River and its harbor at Manhattan, it was wealthier, too.

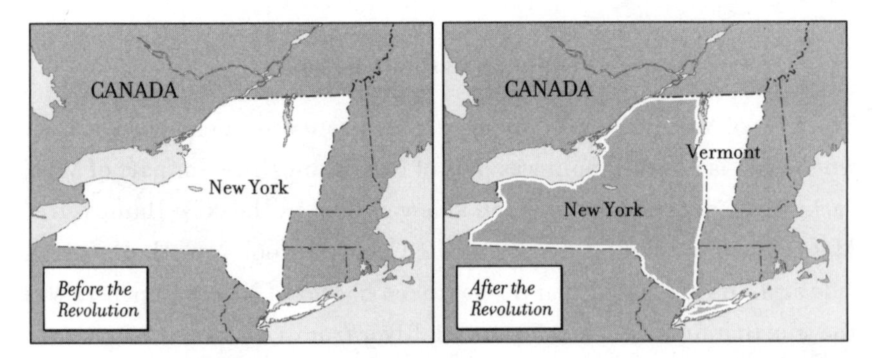

FIG. 165 The Emergence of Vermont

The quest for equality, however, was not a characteristic of British rule. From the king's point of view, New York was large, affluent, generally content, and not much motivated to venture into Canada. New Hampshire was small, less affluent, and expansionistic. Already it had begun issuing titles to the disputed land west of the Connecticut River. By granting to New York all the land up to the Connecticut River, the king knew any moves toward Quebec by New Hampshire would be thwarted by the highlands in its north and by the powerful colony of New York on its west. (See Figure 113, in NEW HAMPSHIRE.)

Much of this changed after the Revolution, but one element that did not was the border along the 45th parallel, which sought to separate the Quebecois who had settled south of the St. Lawrence from the Americans. This line remains the northern border of Vermont to this day.

Vermont's Western Border

For all the wisdom embodied in George III's northern border of New York, his eastern border along the Connecticut River was completely unsuccessful. When the New York authorities attempted to tax the

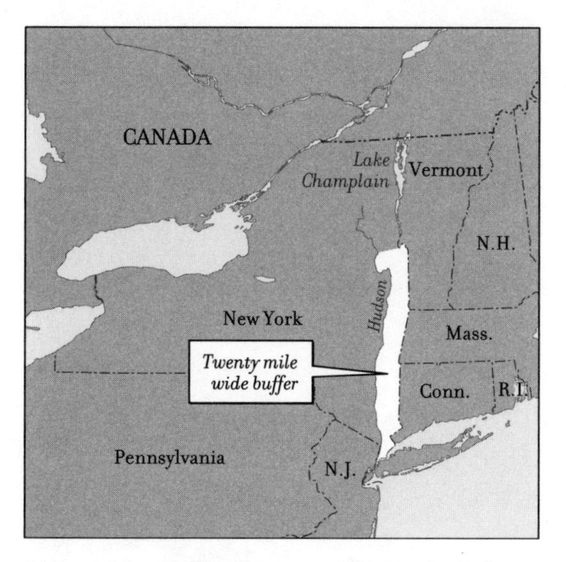

FIG. 166 Hudson River—20-Mile Border Buffer

residents, they found themselves facing the muskets of Ethan Allen and the Green Mountain Boys.

Before violence erupted, however, the American Revolution commenced. In keeping with the spirit of the time, the region declared its own independence as the state of Vermont. When the Continental Congress refused to recognize it, Vermont threatened to ally itself with England. In response, Congress voted to invade Vermont! But George Washington resisted, pointing out that his troops had little desire to fight fellow Americans.

Hence, in 1789, Congress recognized the state of Vermont, and to divide it from New York, Congress established a border that continued the 20-mile-wide buffer used to separate the Hudson from Massachusetts and Connecticut. (Figure 166) Beyond the Hudson, Lake Champlain served as an ideal continuation of the line.

VIRGINIA

Why does Virginia's northern border depart from the Potomac so weirdly? Why is Virginia's straight-line southern border located where it is? And how did Virginia get the bottom part of Maryland's eastern shore? It's not even connected to Virginia!

Virginia was the first British colony established in America. In 1606, King James I issued its charter, according to which Virginia's boundaries encompassed the land between 34° and 41° N latitude from the

Atlantic Ocean to the Pacific Ocean—a swath of land that crosses the continent starting from what is now the southeast corner of North Carolina and the northeast corner of New Jersey. James further expanded the claim in 1611 by extending Virginia's southern border to the 29°, which is where Daytona Beach is today.

After that, the boundaries of the Virginia Colony began to recede as the crown issued charters for other colonies. When King Charles I issued the charter creating Maryland, the first of Virginia's present-day borders surfaced.

Virginia's Northern Border

The creation of Maryland in 1632 resulted in Virginia acquiring its present northern border, the Potomac River. It also resulted in Virginia acquiring its first border dispute. The Virginia/Maryland boundary was stipulated as the Potomac River, from its juncture with the Chesapeake Bay to its headwaters. But in its upstream reaches, the Potomac branches, raising the question: Which branch is the main branch? (See Figure 90, in MARYLAND.) Virginia took the position that the northern branch was the main branch, not simply because it resulted in more land for Virginia but because Virginia, having previously encompassed this land, had already issued deeds to its owners. Maryland argued for the south branch, based on its being the larger branch, but ultimately Virginia's position was upheld.

Virginia's Eastern Border

Prior to the creation of Maryland, Virginia's plantations had spread out from the colony's original settlements at Jamestown and Williamsburg, with some colonists crossing the Chesapeake Bay and cultivating the land at the southern end of the peninsula between the Chesapeake and the ocean. (See Figure 89, in MARYLAND.) Charles I envisioned this peninsula as being part of the Maryland colony he was creating. But since

Virginia, having until now encompassed this land, had deeded its properties, Maryland's charter states that the Virginia/Maryland border on this peninsula is to be a line from Watkin's Point due east to the ocean.

As it turned out, the line that was surveyed did not begin at Watkin's Point, since Watkin's Point had eroded away. But a memory of its location remained, though not, evidently, in the minds of the surveyors. They commenced their line several miles north of where Watkin's Point had been. Compounding the problem, they then proceeded to mark off a line that veered even farther to the north. Maryland contested the line for over a hundred years, but ultimately its claim was denied and the line, though faulty, remains to this day. (For more details, go to MARYLAND.)

Virginia's Southern Border

In 1663, King Charles II revised Virginia's borders yet again, when he created the Carolina Colony. The Carolina Colony's northern border was an east-west line located at 36°, the middle of Albemarle Sound, and extending from ocean to ocean. But Virginia continued its practice of charging a tariff on the goods entering and departing from Albemarle Sound, regardless of whether the ships were doing business with Virginia or Carolina. Carolina objected to this practice and the king agreed. In 1665, Charles II relocated the Virginia/Carolina border one-half of a degree farther north. This placed it midway between Albemarle Sound and the Chesapeake Bay, which is where Virginia's southern border remains to this day. (See Figure 125, in NORTH CAROLINA.)

The king's adjustment had an enormous impact in the future. The border between Virginia and what is now North Carolina extends, with some dips and interruptions, all the way through the top of the Texas panhandle. Most significantly, it formed the basis for the Missouri Compromise (1820), a key event in the American effort to avert a war over slavery. (For more, see DON'T SKIP THIS.)

For all the influence this border at 36°30' has had, ironically it is mislocated. At its starting point, the boundary was erroneously located (and

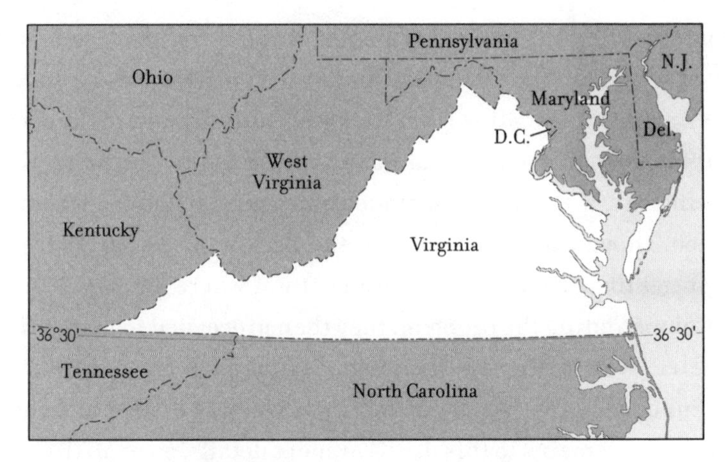

FIG. 167 **The Northward Deviation of Virginia's Southern Border**

remains to this day) slightly north of 36°30'. At the western end of Virginia, the segment surveyed by Thomas Jefferson's father, Peter Jefferson, veers farther to the north. (Figure 167) By the time the line arrives at the southwestern corner of Virginia, it is actually over 5 miles north of 36°30'.

Virginia's Western Border

Virginia's western border was considerably foreshortened following the French and Indian War (1754–63). France's defeat in this war resulted in France's relinquishing to England all its land claims between the Ohio River and the Mississippi River. England, fearing its American colonists might expand their settlements beyond England's control, prohibited the Americans from migrating beyond the Ohio River. In effect, Virginia's western border was now the Ohio River.

England's prohibition outraged the Americans, who had, after all, fought side by side with the British in the war. England sought to mollify the colonists by allowing them to form corporations to engage in land speculation along the Ohio frontier.

After the Revolution, the new United States government urged those states with extensive land claims to donate that land to the federal government. Virginia released some of its western lands but not to the land west of the Appalachians that bordered its investments along the Ohio River. Today, this land is known as West Virginia, but back then it was Virginia—although Pennsylvania claimed the northern part.

This dispute resulted from uncertainty as to the western end of Pennsylvania. (For more on that, go to PENNSYLVANIA.) It was a particularly violent dispute, known as Dunmore's War, because at issue was the juncture of the Monongahela and Allegheny rivers, which together form the Ohio River. (Today, we call this juncture Pittsburgh.) Virginia laid claim to the region based on its royal charters and its interpretation of Pennsylvania's charter. (See Figure 146, in PENNSYLVANIA.) Ultimately, the Continental Congress ruled on the location of Pennsylvania's western border, giving Pittsburgh to Pennsylvania but enabling Virginia to keep virtually all of the land in which it had invested along the Ohio.

Virginia then released its remaining western land claims, thereby creating Kentucky. To separate itself from Kentucky, Virginia created a border that followed the crest of the Appalachian Mountains northeastward, then crossed a series of valleys along a straight northeast line until arriving at the Tug Fork River. Here the border turns to the northwest, following the Tug Fork to the Big Sandy River to the Ohio River. (Figure 168) This border enabled the western region of Virginia to have access, via Tug Fork and the Big Sandy, to the Ohio River and onward to the Mississippi River and the sea.

Virginia's western border acquired its final shape when its mountainous region separated itself and became West Virginia. The land in this section of the state was unsuited to large plantations, and the farmers who eked out an existence on that land could barely afford horses, let alone slaves. Yet the apportionment of representatives to the state legislature was based on population counts that included slaves, even though the slaves could not vote. Consequently, the legislature rarely distributed state services equally between its eastern and western regions. As early as the 1820s, Virginia's mountain population began to speak of

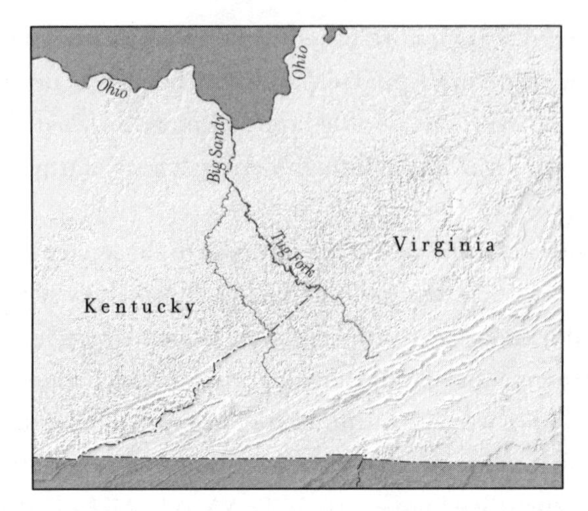

FIG. 168 Virginia's Creation of Kentucky—1790

forming a separate state. With the onset of the Civil War, an ideal opportunity presented itself. After Virginia voted to secede from the Union, its mountain districts voted to secede from Virginia.

The location of the boundary that Congress fixed between Virginia and West Virginia reflects which counties were under the protection of (some would say occupied by) Union troops in October 1861, when the citizens voted to create a new state. (See Figure 173, in WEST VIRGINIA.) Congress later appropriated three Shenandoah Valley counties that had not voted to join West Virginia. These counties account for the unusual eastern end of West Virginia, and thus the odd way Virginia departs from the Potomac.

Of course, in wartime boundaries are meaningless. After the war, however, they were not. Virginia protested what it viewed as the unconstitutional alteration of its border. Proponents of West Virginia argued that Virginia, having seceded from the Union, could not now claim protection under the Constitution. Virginia countered that the federal government's claim that a state could not secede was the reason for the Civil War and, therefore, the federal government could not abrogate Virginia's constitutional rights.

Virginia may have been right but, in the wake of so much bloodshed, it could not win the sympathy of the U.S. Supreme Court. Today, the controversial border between Virginia and West Virginia preserves the history of an era of moral and legal controversy, as well as a very specific moment from that era when certain regions in western Virginia were occupied by the Union army.

WASHINGTON

Why does the southern border of Washington suddenly turn into a straight line? And why does it stop where it does? How come Idaho has that thin strip in between Washington and Montana rather than Washington just continuing to Montana?

Washington's Northern Border

By the time Congress decided it was time to create the state of Washington, many of its boundaries were already in place. The oldest of those boundaries was its northern border, which follows the 49th parallel. This boundary first surfaced in 1818 but nowhere near the state of Washington. It was above Minnesota. (To find out why—and why they chose the 49th parallel—see DON'T SKIP THIS.) In 1846, England and the United States decided to separate their interests in the Oregon Country, a vast area that they jointly held. (See Figure 142, in OREGON.) The two nations agreed to extend their boundary at the 49th parallel across the Rocky Mountains all the way to Puget Sound on the Pacific coast. (For more on this agreement and the threat of war that surrounded it, go to OREGON.) This treaty resulted in the creation of the Oregon Territory. The northern border of the Oregon Territory later became the northern border of the state of Washington and has remained so to this day.

Washington's Southern Border

The southern border of Washington was established in 1853, when residents of the Oregon Territory who lived north of the Columbia River argued that the distance to the territorial capital (Oregon City, in the Willamette Valley) was so great that government services were too difficult to access. Congress responded by dividing the Oregon Territory horizontally along the Columbia River, from its mouth at the Pacific Ocean to the point where its flow first crosses the 46th parallel. From this point, the border follows the 46th parallel to the crest of the Rocky Mountains—the eastern extent, at that time, of the Oregon Territory. (To find out why Washington and Oregon weren't simply divided by a horizontal line midway between them, go to OREGON.)

But why stop following the river at the 46th parallel? Indeed, when Oregon applied for statehood in 1857, it proposed that its border with Washington be the Columbia River from the Pacific to the point at which the Columbia is joined by the Snake River, and that the Snake River then take over as the border all the way to its source at the Continental Divide. (Figure 169)

At first glance, it doesn't seem like such a bad idea. But Washington's

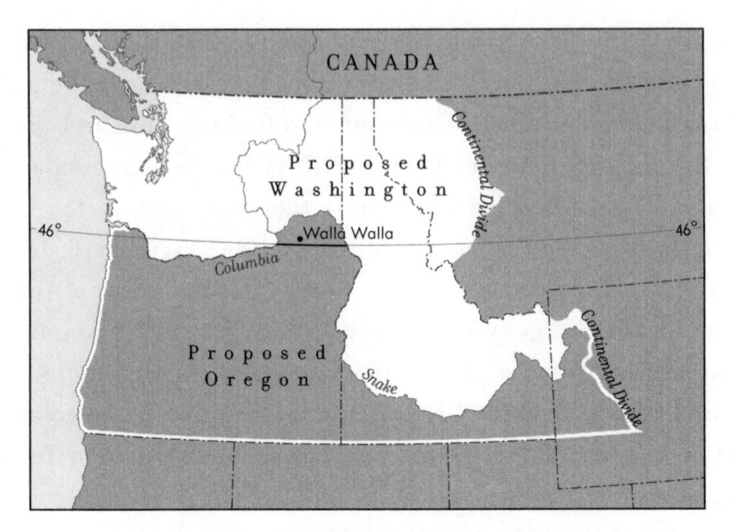

FIG. 169 The Proposal to Divide Washington and Oregon by Rivers

territorial governor Fayette McMullen termed it a "political, moral, and social outrage." What was he so steamed up about? The answer can be summed up in one word, said twice: Walla Walla.

Walla Walla is a Native American term meaning "many waters." But it wasn't the settlement at Walla Walla whose potential loss aroused the governor's ire. It was the many streams that converged in this fertile region, a rarity in the arid eastern half of the Oregon Territory. Dividing Washington from Oregon using the Snake River rather than the 46th parallel places this region in Oregon. For Oregon, with its large and productive Willamette Valley, to obtain this additional pocket of productive land was the "political, moral, and social outrage" to which Governor McMullen was referring.

Washington's Eastern Border

Washington's eastern border has had the most interesting history, at least from a visual point of view. When Oregon became a state in 1859, the land left over by its newly defined eastern border was annexed to the

Washington Territory, resulting in Washington having a rather peculiar shape. (Figure 170) But not for long.

In 1860, gold was discovered in the eastern mountains of the Washington Territory. Suddenly, the territory found itself home to a flood of new residents, many of them coming from backgrounds quite different from those of the people already settled in the agricultural areas. In order to preserve their political control, Washington's residents agreed to separate from the gold-mining region—a separation that would have come anyway, as evidenced by the eastern boundary Congress selected for Washington.

The eastern border of Washington is composed of a continuation of the same north-south segment of the Snake River that borders the eastern end of Oregon. Also mirroring the eastern border of Oregon, the Snake River is replaced with a straight north-south line upon reaching its juncture with another river, in this instance the Clearwater River. And just as in the case of Oregon, that straight line is located at 117° W longitude. (Figure 171)

With these eastern borders, Washington and Oregon have seven

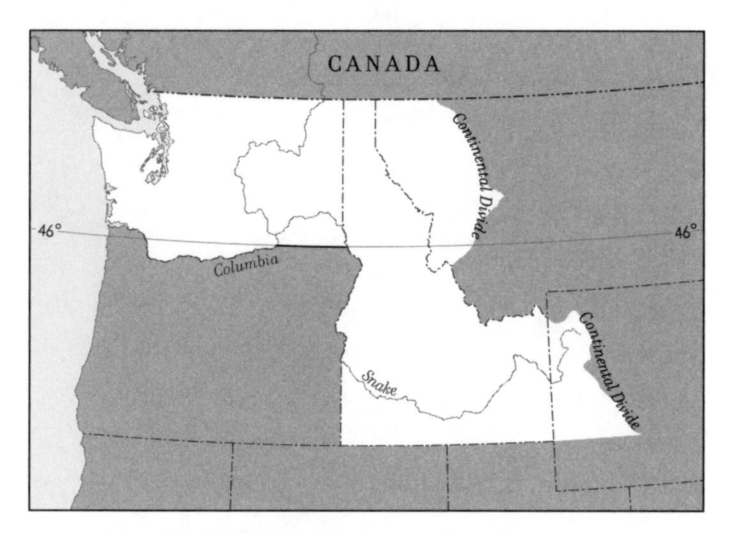

FIG. 170 **The Washington Territory—1859–1862**

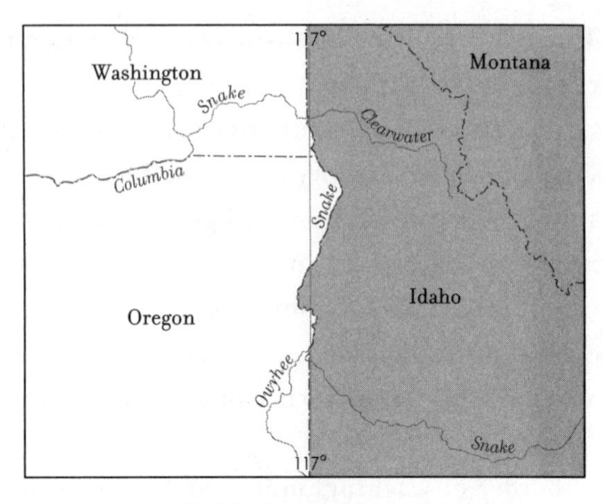

FIG. 171 Washington and Oregon's Eastern Border

degrees of width (more or less, given their ocean borders)—the same width as North Dakota, South Dakota, Wyoming, and Colorado. Thus, in the eastern border of Washington we discover yet another artifact of the policy to which one Congress after another has continued to subscribe: All states should be created equal.

WEST VIRGINIA

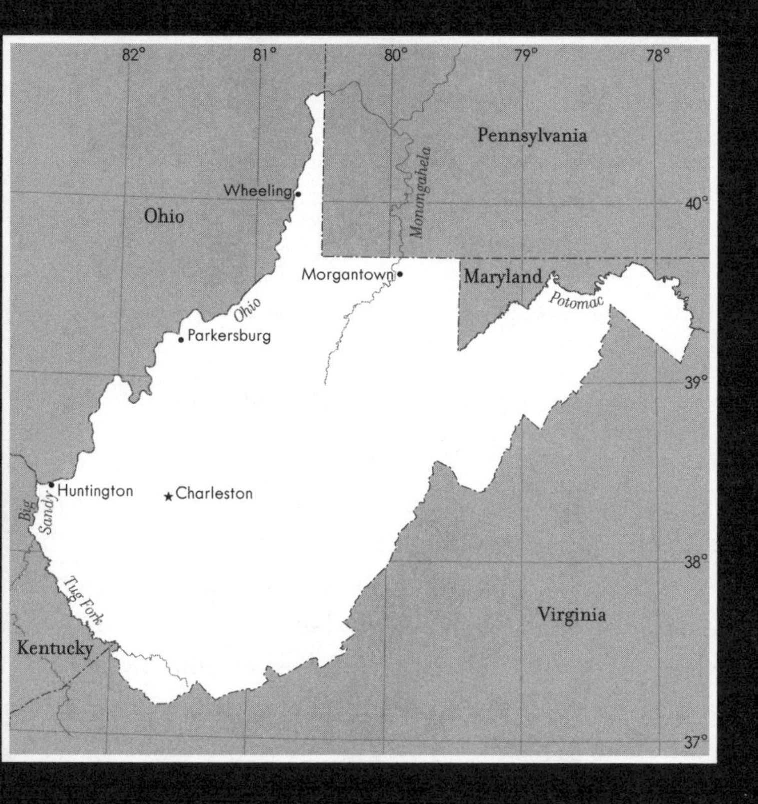

Why does West Virginia have that skinny little part crawling up the side of Pennsylvania? Why does it have that other weird thing coming up underneath Maryland?—wouldn't it be better if that western triangle of Maryland was part of West Virginia? And how come there isn't East Virginia?

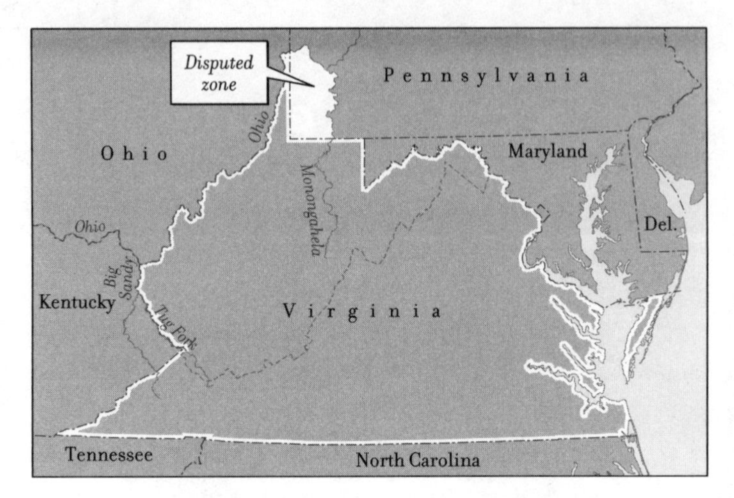

FIG. 172 Beginnings of West Virginia Border—1790

West Virginia's Western Borders

The area that we call West Virginia was previously part of (no surprise here) Virginia. After the Revolution, Congress urged those states with vast colonial boundaries to donate their extended regions to the United States. Virginia released its claim to what is today Kentucky, in addition to all its lands beyond the Ohio River. But in order to maintain access to its recent investments in land along the upper Ohio River, Virginia retained the land leading to that segment of the river. Virginia therefore located the middle of what was then its western border along the Tug Fork River and the Big Sandy River. Virginia's efforts are preserved today in the western border of West Virginia. (Figure 172)

West Virginia's Northern Border

Virginia's land investments along the Ohio River profoundly affected the northern border of what would become West Virginia. Because of overlapping land claims, Virginia and Pennsylvania became embroiled in a boundary dispute in the upper Ohio River region. The dispute led to Pennsylvania's border being located just west of Pittsburgh, where the Ohio River begins. This left Virginia—and, years later, West Virginia—with a section of its western border being a thin finger of land between the Ohio River and Pennsylvania. (For more on this, go to PENNSYLVANIA.)

West Virginia's Eastern Border

Those who lived in the western end of Virginia were not the happiest of citizens. Their rocky and mountainous terrain could not produce nearly as abundant a crop as that of Virginia's Piedmont and Tidewater regions. The lack of affluence that resulted precluded farmers in the region from making the costly investment of owning slaves. To make matters worse, representation in the Virginia legislature was apportioned by a census in which slaves were included in the count (though, of course, not in the right to vote). Thus the Tidewater and Piedmont regions added greater power to their greater wealth relative to that of the state's Appalachian residents. On several occasions, the citizens of western Virginia asked the legislature to allow their region to become a separate state, but to no avail.

With the Civil War, however, the residents of western Virginia saw their chance. Since Virginia had declared its right to secede from the Union, the state's western residents seceded from Virginia. That opportunity presented itself in 1861, when Union troops crossed the Ohio River and attacked Confederate defenses in the mountains of what was then western Virginia. Numerous battles and skirmishes followed throughout 1861. (Figure 173) By autumn, the Confederate lines had withdrawn southeasterly. In October, western Virginians voted to join the Union as a separate state. The border between West Virginia and Virginia preserves that moment in time when federal troops were present in those counties that are now West Virginia.

Congress also added three nearby counties to West Virginia—Morgan, Berkeley, and Jefferson—despite the fact that they were not among those that voted. Indeed, these Shenandoah Valley counties were loyal to Virginia. But they were located in very fertile farmland, through which the Baltimore and Ohio Railroad ran. The B&O connected Baltimore and Washington (and now the state of West Virginia) to the Ohio River and, via the Ohio, the Mississippi River and the sea. (Figure 174)

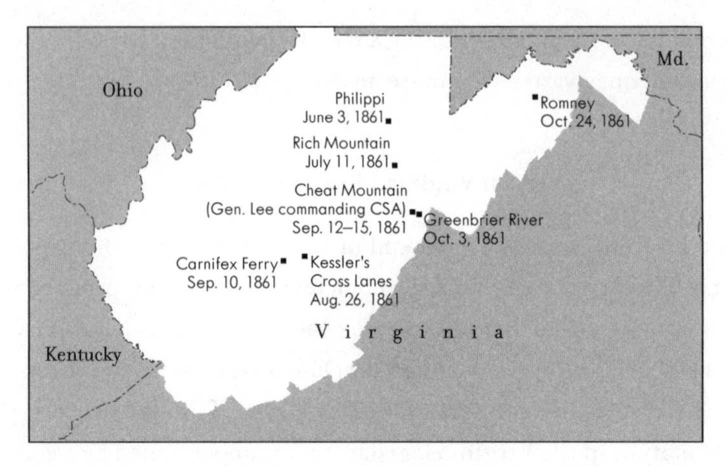

FIG. 173 Troop Locations Leading to West Virginia Statehood Vote

Although the Shenandoah Valley changed hands many times during the Civil War, Congress annexed these counties to West Virginia in an effort to provide it with the resources needed to sustain itself as a state. Virginia sued for the return of these counties in 1871. But having only recently seceded from the Union, and having lost the war, Virginia was not in a very strong bargaining position. The U.S. Supreme Court ruled against Virginia's claims.

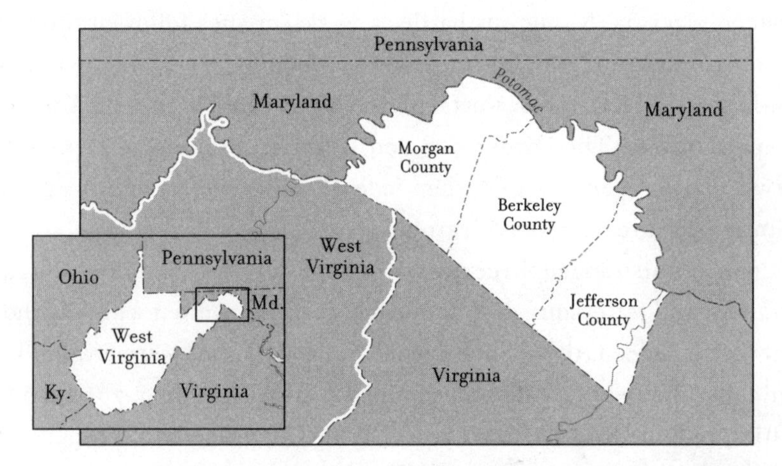

FIG. 174 Counties Later Added to West Virginia

WISCONSIN

How come the peninsula that extends off the northeast corner of Wisconsin is in Michigan? Why is Wisconsin's straight-line southern border located where it is? Wouldn't it look nicer if it lined up with Michigan's southern edge across the lake?

What we now call Wisconsin was previously part of the Northwest Territory, land that the British and their American colonists won from France in the French and Indian War, 1754–1763. (See Figure 4, in DON'T

SKIP THIS.) After the American Revolution, Congress enacted the Northwest Ordinance (1787), which contained boundary lines for the eventual division of the region into states. Wisconsin was the last of these states to acquire sufficient population for statehood. The fact that it was last significantly impacted its borders.

Wisconsin's Southern Border

Illinois, upon becoming a state in 1818, successfully argued for the acquisition of over 60 miles of what would have been southern Wisconsin. (Figure 175) These sixty miles were a continuation of the relatively flat land that characterized most of northern Illinois. Illinois' founders sought to include this land in a system of canals they envisioned that would alter the flow of commerce in northern Illinois away from the Mississippi River (and the slave-holding South) to Lake Michigan and, via the Erie Canal, to New York. Even in 1818, the division between North

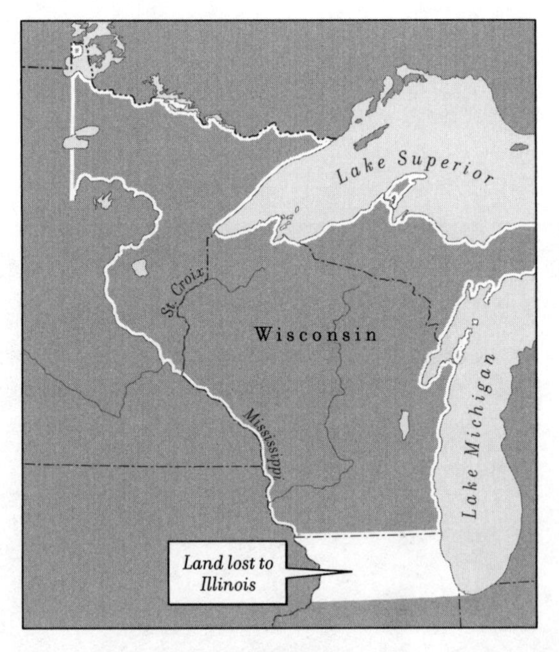

FIG. 175 Wisconsin's Southern Border—1818

and South was of so much concern that a majority in Congress voted to allow this annexation.

Wisconsin's Eastern Border

In 1833, as compensation to Michigan for the land it lost to Indiana and Ohio, Congress gave Michigan the Upper Peninsula of Wisconsin. (Figure 176. For more details, go to OHIO and INDIANA.) Because this act by Congress ended a threat of genuine violence (remembered in history as the Toledo War), Wisconsin knew it could not successfully protest.

Wisconsin did, however, stand its ground regarding just how much of the Upper Peninsula was given to Michigan. At issue was that portion of the border that Congress declared was to follow the Montreal River "as marked upon the survey made by Captain Cram." (Thomas Cram was the Army engineer assigned to survey the peninsula.) But the Montreal River branches. Since both branches are of equal depth, which is the river? Captain Cram had surveyed along the eastern branch. Michigan

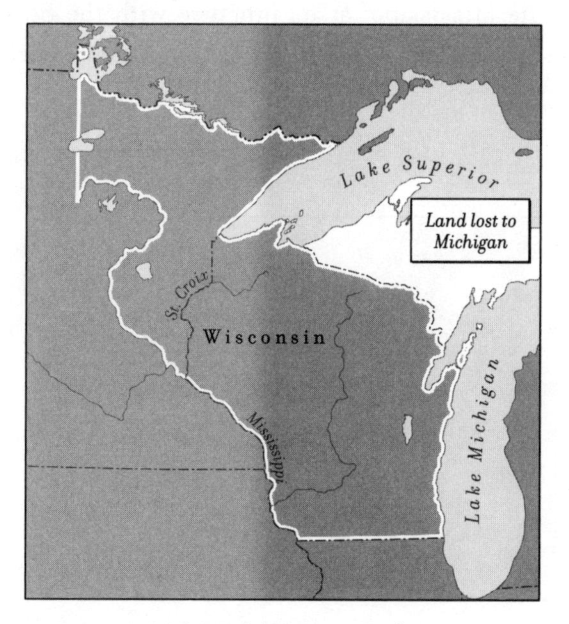

FIG. 176 Wisconsin's Eastern Border—1833

wanted the western branch. Three hundred and sixty square miles hinged on the answer. (Figure 177)

When Michigan revised its Constitution in 1908, it described this section of its boundary as "the westerly branch of the Montreal River." Two states were now officially claiming jurisdiction over the same land. To rectify the issue, Michigan sent negotiators to Wisconsin. But they left without so much as a souvenir cheese. As far as Wisconsin was concerned, the eastern branch identified by Captain Cram was the border. Though never officially resolved, the eastern branch continues to be recognized as the peninsular border of Wisconsin.

Wisconsin's Western Border

When defining what would become the final borders for the new state of Wisconsin, Congress sliced off from its western region one last hunk of land, which it then used to create the state of Minnesota. (Figure 178) The Northwest Ordinance had stipulated that Wisconsin's western border would follow the Mississippi River. The new border, however, departed from the Mississippi at its juncture with the St. Croix River, following the St. Croix to its point due south of the western tip of Lake

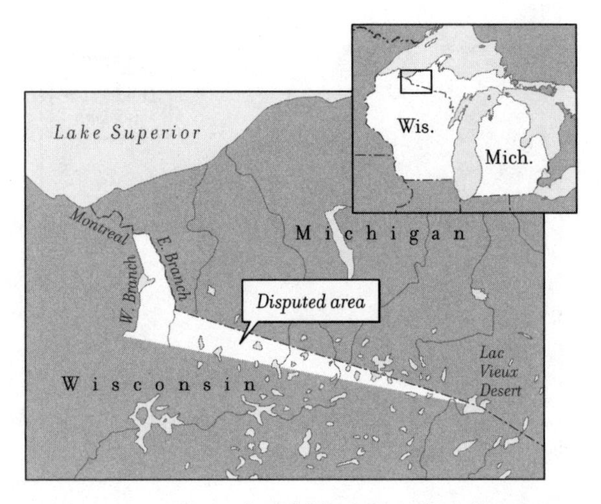

FIG. 177 Wisconsin/Michigan Boundary Dispute

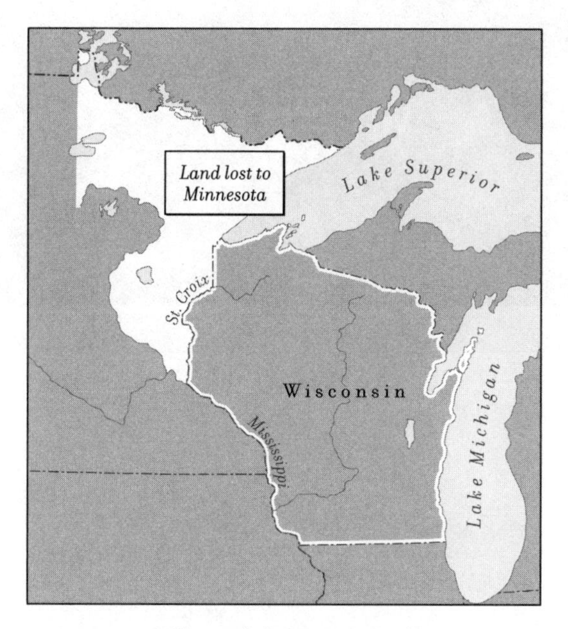

FIG. 178 Wisconsin's Western Border—1846

Superior. A straight line then joined that point in the river with the western end of Lake Superior. From there, the lake then served as the northern border of Wisconsin up to the Montreal River.

In order for the states in this region to be as equal as possible, each needed access to the Great Lakes. With Wisconsin's original borders, the state that would eventually be created to its west would not have had a window on any of the Great Lakes. But with the 1846 border adjustment, the future state to the west of Wisconsin, Minnesota, gained access to Lake Superior, while Wisconsin retained access not only to Lake Superior, but to Lake Michigan as well.

WYOMING

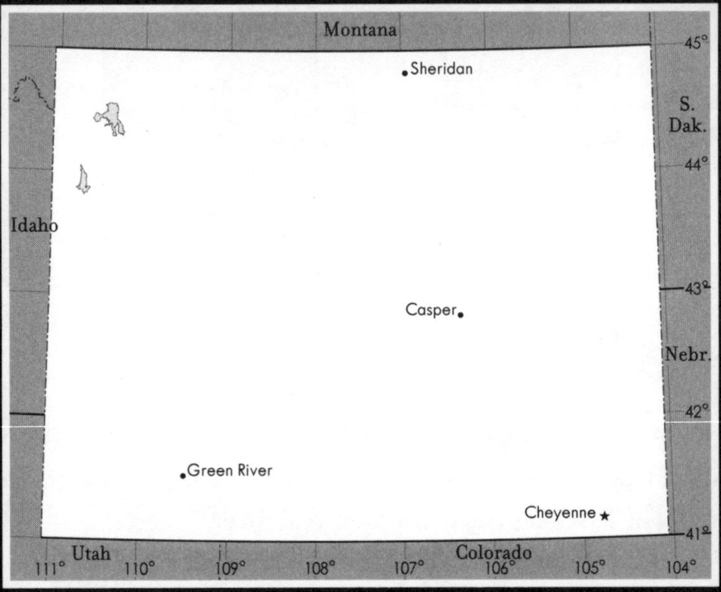

How come Wyoming's southwest corner takes a bite out of Utah rather than the other way around? Why are the lines of Wyoming's borders located where they are? Couldn't they have lined up east and west with Colorado's?

Wyoming's Southern and Northern Borders

When Wyoming became a territory in 1868, three of its current borders were already in place. The northern border of Colorado provided the

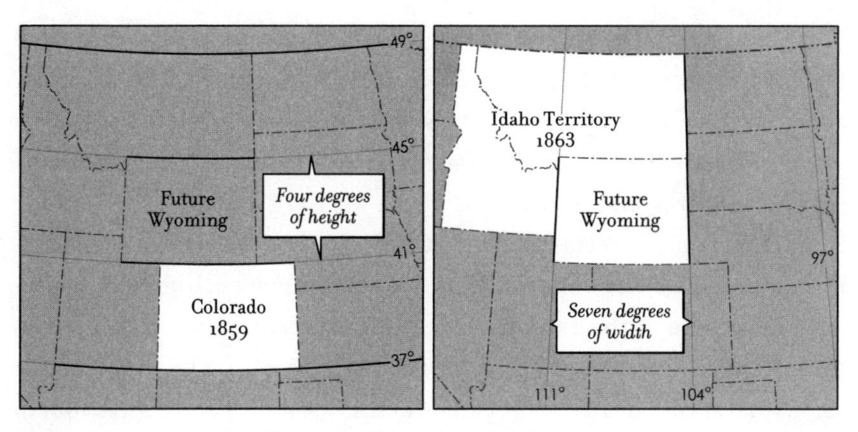

FIG. 179 Formative Elements of Wyoming's Borders

southern border for what was to become Wyoming. Colorado also provided Wyoming with its northern border, despite the fact that Wyoming's northern border is not even close to Colorado.

When Colorado applied for territorial status in 1859, Congress adjusted its proposed northern and southern borders. This adjustment resulted in Colorado having a height equal to one-third the distance between its southern border and the Canadian border. Thus, two more states of equal height could fit between Colorado and Canada. Some years later, Montana, with precisely the same height as Colorado, was established as a territory, and thereby the northern border of what would then become Wyoming was now in place. (Figure 179)

Wyoming's Eastern and Western Borders

Wyoming's eastern border with Nebraska and the Dakotas was also already in place when it became a territory. Back in 1863, when Congress had created the original Idaho Territory, its eastern border was established at 104° W longitude. The Idaho Territory was soon divided into two territories, but 104° remained as a border dividing the Dakotas and Nebraska, on the east side of this line, from Montana and Wyoming, on the west side.

But why 104°? The reason revealed itself when Congress later located Wyoming's western border. This line, at 111° W longitude, gave Wyoming seven degrees of width. As such, Wyoming joined five other western states with equal width: North Dakota, South Dakota, Colorado, Washington, and Oregon. (Figure 179)

That Wyoming's rectangle got to take a corner from Utah's rectangle, rather than the other way around, would certainly seem like an arbitrary decision. Some of the reasons were political. (For more on this, go to UTAH.) But the main reason was based on the terrain. The Uinta Mountains turn at very nearly a right angle at this location. (See Figure 163, in UTAH.) Had Congress apportioned this corner to Utah, it would have resulted in Utah having a valley from which it was entirely blocked by mountains. For Wyoming, on the other hand, access to the valley is direct and the mountains become a natural border.

Selected Bibliography

General

Berkhofer, Robert F., Jr. "Jefferson, the Ordinance of 1784, and the Origins of the American Territorial System," *William and Mary Quarterly*, 3rd Ser., Vol. 29, no. 2 (April, 1972): pp. 243–245.

Kappler, Charles J., ed. *Indian Affairs: Laws and Treaties*. 2nd ed. Washington, D.C.: Government Printing Office, 1904.

Library of Congress, Geography and Map Division. *American Memory: Map Collections, 1500–2002*. Online. http://lcweb2.loc.gov/ammem/gmdhtml/gmdhome.html. (accessed 2003.)

Royce, Charles C. *Indian Land Cessions in the United States*. New York: Arno Press, 1971.

Shearer, Benjamin F., ed. *The Uniting States: The Story of Statehood for the Fifty United States*. (Westport, Connecticut: Greenwood Press), 2004.

Shepherd, William R. *Historical Atlas*. 9th ed. New York: Henry Holt & Co., 1964.

United States Geological Survey. *National Atlas of the United States.* Online http://nationalatlas.gov/index.html (accessed 2007.)

Van Zandt, Franklin K. *Boundaries of the United States and the Several States*. Washington, D.C.: Government Printing Office, 1976.

Yale University. *The Avalon Project at Yale Law School: Documents in Law, History, and Diplomacy.* Online. http://www.yale.edu/lawweb/avalon/avalon.htm. (accessed 2003.)

Alabama

Abernethy, Thomas Perkins. *The Formative Period in Alabama: 1815–1828.* Montgomery, AL: Brown Printing Co., 1922.

Pickett, Albert James. *History of Alabama.* Charleston, SC: Walker & James, 1851.

Alaska

Harrison, John Armstrong. *The Founding of the Russian Empire in Asia and America.* Coral Gables, FL: University of Miami Press, 1971.

Naske, Claus M. *Alaska: A History of the 49th State.* Norman, OK: University of Oklahoma Press, 1987.

Arizona

Bancroft, Hubert H. *History of Arizona and New Mexico, 1530–1888.* San Francisco, CA: History Co., 1889.

Walker, Henry P. *Historical Atlas of Arizona.* Norman, OK: University of Oklahoma Press, 1979.

Arkansas

Arnold, Morris S. *Colonial Arkansas: 1686–1804.* Fayetteville, AR: University of Arkansas Press, 1993.

Bolton, S. Charles. *Territorial Ambition: Land and Society in Arkansas, 1800–1840.* Fayetteville, AR: University of Arkansas Press, 1993.

California

Beck, Warren A. *Historical Atlas of California.* Norman, OK: University of Oklahoma Press, 1974.

Lavender, David. *California: A Bicentennial History.* New York: Norton, 1976.

Colorado

Abbott, Carl. *Colorado: A History of the Centennial State.* Boulder, CO: Colorado Associated University Press, 1976.

Connecticut

Bowen, Clarence Winthrop. *The Boundary Disputes of Connecticut*. Boston, MA: James R. Osgood & Co., 1882.

Dwight, Theodore. *The History of Connecticut*. New York: Harper & Bros., 1840.

Delaware

Lunt, Dudley Cammett. *The Bounds of Delaware.* Wilmington, DE: Star Publishing Co., 1947.

Schenck, William S. "Delaware's State Boundaries." Newark, DE: Delaware Geological Survey, 2005. Online. http://www.udel.edu/dgs/Publications/pubsonline/info6.html (accessed May 2006).

District of Columbia

Green, Constance McLaughlin. *Washington*. Princeton, NJ: Princeton University Press, 1962–63.

Florida

Patrick, Rembert Wallace. *Florida Under Five Flags*. Gainesville, FL: University of Florida Press, 1945.

Georgia

Coleman, Kenneth. *A History of Georgia*. Athens, GA: University of Georgia Press, 1977.

Martin, Harold H. *Georgia: A Bicentennial History*. New York: Norton, 1977.

Hawaii

Stevens, Sylvester Kirby. *American Expansion in Hawaii: 1842–1898*. New York: Russell & Russell, 1945.

Thomas, Benjamin E. "Demarcation of the Boundaries of Idaho," *Pacific Northwest Quarterly*, Vol. XL (January 1949).

Idaho

Beal, Merrill D., and Merle W. Wells. *History of Idaho.* New York: Lewis Historical Publishing Co., 1959.

Peterson, Frank Ross. *Idaho: A Bicentennial History.* New York: Norton, 1976.

Illinois

Brown, Henry. *The History of Illinois.* New York: J. Winchester, 1844.

Indiana

Dillon, John B. *The History of Indiana.* Indianapolis, IN: Bingham & Doughty, 1859.

Iowa

Andreas, A. T. *Illustrated Historical Atlas of the State of Iowa.* Chicago, IL: Andreas Atlas Co., 1875.

Dodds, J. S., ed. *Original Instructions Governing Public Land Surveys of Iowa: A Guide to Their Use in Resurveys of Public Lands.* Ames, IA: Iowa Engineering Society, 1943.

Larzelere, Claude S., Harlow Lindley, and Bernard C. Steiner. "The Iowa-Missouri Disputed Boundary." *The Mississippi Valley Historical Review,* (1916): pp. 77–84.

Kansas

Cutler, William G. *History of the State of Kansas.* Chicago, IL: A. T. Andreas, 1883.

Gower, Calvin W. "Kansas Territory and Its Boundary Question: 'Big Kansas' or 'Little Kansas.' " *The Kansas Historical Quarterly,* Vol. 33, no. 1 (Spring 1967).

University of Kansas, ed. *Territorial Kansas: Studies Commemorating the Centennial.* Lawrence, KS: University of Kansas Press, 1954.

Kentucky

Marshall, Humphrey. *The History of Kentucky.* Frankfort, KY: G. S. Robinson, 1824.

Louisiana

Adams, John Quincy. *Diary of John Quincy Adams.* Edited by Allen Grayson et al. Cambridge, MA: Belknap Press of Harvard University Press, 1981.

Brooks, Philip Coolidge. *Diplomacy and the Borderland: The Adams-Onis Treaty of 1819*. Berkeley, CA: University of California Press, 1939.

Jefferson, Thomas. "The Proper Size of States." Online. http://etext.lib.virginia.edu/jefferson/quotations/jeff0400.htm (accessed March 2006).

Labbé, Dolores Egger. *The Louisiana Purchase and Its Aftermath, 1800–1830*. Lafayette, LA: Center for Louisiana Studies, University of Southwestern Louisiana, 1998.

Laussat, Francaise. *"Proclamation au nom de la republique,"* broadside (New Orleans, 1802). Library of Congress Printed Ephemera Collection; Portfolio 24, Folder 5. (Digital id: [rbpe 02400500] http://hdl.loc.gov/loc.rbc/rbpe.02400500).

Wall, Bennett H., and Light T. Cummins, eds. *Louisiana: A History*. 4th ed. Wheeling, IL: Harlan Davidson, 2002.

Maine

Abbott, John S. *The History of Maine*. Boston, MA: B. B. Russell, 1875.

Maryland

Papenfuge, Edward, and Joseph Coale IV. *Atlas of Historical Maps of Maryland: 1608–1908*. Baltimore, MD: Johns Hopkins University Press, 1982.

Massachusetts

Bradford, Alden. *History of Massachusetts for Two Hundred Years: From the Year 1620 to 1820*. Boston, MA: Hilliard, Gray & Co., 1835.

Michigan

Utley, Henry M. *Michigan as a Province, Territory, and State: The Twenty-sixth Member of the Federal Union*. New York: Publishing Society of Michigan, 1906.

Minnesota

Butler, Nathan. *Boundaries and Public Land Surveys of Minnesota*. [no date or publisher cited]

Davis, Samuel. *The Dual Origin of Minnesota*. [no publisher cited] 1901

Folwell, William W. *A History of Minnesota*. St. Paul, MN: Minnesota Historical Society, 1921–1930.

Squires, Rod. "Minnesota Boundary Line." Online. http://www.geog.umn.edu/faculty/squires/research/RealProp/survey/mnboundaries.html (accessed March 2006).

Mississippi

Clayborne, John Francis Hamtramck, and C. M. Lagrone. *Mississippi as a Province, Territory, and State.* Jackson, MS: Power & Barksdale, 1880.

Missouri

Ellis, James Fernando. *The Influence of Environment on the Settlement of Missouri.* St. Louis, MO: Webster Publishing Co., 1929.

March, David. *History of Missouri*. Vol. I. New York: Lewis Historical Publishing Co., 1967.

Montana

Deutsche, Herman J. "The Evolution of State Boundaries in the Inland Empire of the Pacific Northwest." *Pacific Northwest Quarterly*, Vol. LI, (July 1960).

Malone, Michael P., and Richard B. Roeder. *Montana: A History of Two Centuries.* Seattle, WA: University of Washington Press, 1976.

Nebraska

Olson, James C. *History of Nebraska.* Lincoln, NB: University of Nebraska Press, 1966.

Nevada

Angel, Myron. *History of Nevada.* 1881. New York: Arno Press, 1973.

New Hampshire

Belknap, Jeremy. *The History of New Hampshire*. Boston, MA: Bradford & Read, 1813.

Ramsdell, George A. *The History of Milford.* Concord, NH: Rumford Press, 1901.

New Jersey

Gordon, Thomas Francis. *The History of New Jersey.* Trenton, NJ: D. Fenton, 1834.

New Mexico

Haines, Helen. *History of New Mexico.* New York: New Mexico Historical Publishing Co., 1891.

Loyola, Mary. *The American Occupation of New Mexico, 1821–1852.* Albuquerque, NM: University of New Mexico Press, 1939.

"*Memorial, al honorable congreso de los Estados Unidos,*" broadside (undated). Library of Congress Printed Ephemera Collection; Portfolio 101, Folder 15. (Digtal id: [rbpe 10101500] http://hdl.loc.gov/loc.rbc/rbpe.10101500).

Reséndez, Adrés. "National Identity on a Shifting Border: Texas and New Mexico in the Age of Transition, 1821–1848." *The Journal of American History* (September 1999): pp. 77–84.

New York

Mather, Joseph H. *A Geographical History of the State of New York.* Utica, NY: H. H. Hawley & Co., 1848.

Sullivan, James, ed. *The History of New York State, 1523–1927.* New York: Lewis Historical Publishing Co., 1927.

North Carolina

Arthur, John Preston. *Western North Carolina: A History, 1730–1913.* 1914. Spartanburg, SC: Reprint Co., 1973.

Ashe, Samuel A. *History of North Carolina.* 1908–1925. Spartanburg, SC: Reprint Co., 1971.

Bamman, Gale Williams. "This Land is Our Land! Tennessee's Disputes with North Carolina." *Genealogical Journal*, Vol. 24, Number 3. Salt Lake City, UT: Utah Genealogical Association (1996).

Byrd, William. *Histories of the Dividing Line Betwixt Virginia and North Carolina.* Introduction and notes by William K. Boyd. New York: Dover Publications, 1967.

North Carolina State Department of Archives and History. *The Formation of the North Carolina Counties, 1663–1943.* Raleigh, NC: State of North Carolina, 1950.

North Dakota

Lounsberry, Clement A. *Early History of North Dakota*. Washington, DC: Liberty Press, 1919.

Ohio

Rosebloom, Eugene H. *The History of the State of Ohio*. Columbus, OH: Ohio State Archaeological and Historical Society, 1941–44.

Oklahoma

Morris, John Wesley, Charles R. Goins, and Edwin C. McReynolds. *Historical Atlas of Oklahoma*. Norman, OK: University of Oklahoma Press, 1942.

"North Fork of the Red River," *The Handbook of Texas Online*. http://www.tsha.utexas.edu/handbook/online/articles/view/NN/rnn8.html (accessed October 31, 2002).

Wardell, M. L. "Southwest's History Written in Oklahoma's Boundary Story: Struggle for Control of Mississippi Valley Leaves Its Mark on State." *Chronicles of Oklahoma*, Vol. 5, no. 3 Oklahoma City, OK; Oklahoma Historical Society (September 1927).

Oregon

Bancroft, Hubert H. *History of Oregon*. San Francisco, CA: History Co., 1886–88.

Pennsylvania

Dunaway, Wayland Fuller. *A History of Pennsylvania*. Englewood Cliffs, NJ: Prentice Hall, 1948.

Long, John H. "The Struggles for Pennsylvania's Boundaries." *Southwestern Pennsylvania Genealogical Society Quarterly*, Vol. 26 (1999).

Russ, William A. *How Pennsylvania Acquired Its Present Boundaries*. University Park, PA: Pennsylvania State University, 1966.

Rhode Island

Rhode Island State Planning Board. *Rhode Island Boundaries, 1636–1936*. Providence, RI: Rhode Island Tercentenary Commission, 1936.

South Carolina

Ramsay, David. *History of South Carolina From Its First Settlement in 1670 to the Year 1808* (1858). Spartanburg, SC: Reprint Co., 1959–60.

South Dakota

Schnell, Herbert S. *History of South Dakota*. Lincoln, NB: University of Nebraska Press, 1975.

Tennessee

Dykeman, Wilma. *Tennessee: A Bicentennial History*. New York: Norton, 1975.

History of Tennessee from the Earliest Time to the Present. Nashville, TN: Goodspeed Publishing, 1887.

Price, Jerry K. *Tennessee History of Survey and Land Law*. Kingsport, TN: Price Publishing, 1976.

Texas

Brown, John Henry. *History of Texas from 1685 to 1892*. St. Louis, MO: L. E. Daniell, 1893.

Stephens, A. Ray, William Holmes, and Phyllis M. McCaffree. *Historical Atlas of Texas*. Norman, OK: University of Oklahoma Press, 1989.

Supreme Court of the United States, *United States v. State of Texas*, 162 U.S. 1, 3 (March 16, 1896).

Utah

Bancroft, Hubert H. *History of Utah, 1540–1886*. San Francisco, MA: History Co., 1889.

Brightman, George F. "The Boundaries of Utah." *Economic Geography*, Vol. 16 (January 1940): pp 87–95.

Vermont

Carpenter, William Henry, and T. S. Arthur. *The History of Vermont*. Philadelphia, PA: Lippincott, Grambo & Co., 1853.

Virginia

Bailey, Kenneth P. "The Ohio Company of Virginia and the Westward Movement, 1748–1792," in Kenneth P. Bailey, *The History of the Colonial Frontier.* Glendale, CA: Arthur H. Clark Co., 1939.

Beverly, Robert. *The History and Present State of Virginia*. Chapel Hill, NC: University of North Carolina Press, 1947.

Washington

Avery, Mary Williamson. *Washington: A History of the Evergreen State.* Seattle, WA: University of Washington Press, 1965.

West Virginia

Rice, Otis K. *West Virginia: A History.* Lexington, KY: University of Kentucky Press, 1985.

"West Virginia Statehood." Online. http://www.wvculture.org/history/statehoo.html (accessed February 2006).

Wisconsin

Kellogg, Louise P. "The Disputed Michigan-Wisconsin Boundary." *Wisconsin Magazine of History*, Vol. 1 (1917–18): pp. 304–307.

Thompson, William Fletcher. *The History of Wisconsin*. Madison, WI: State Historical Society of Wisconsin, 1973.

Wyoming

Larson, T. A. *History of Wyoming.* Lincoln, NB: University of Nebraska Press, 1965.

Index

Page numbers in **bold** refer to main entries of states; numbers in *italics* refer to figures.